T0075091

Laboratory Experiments in Trace Environmental Quantitative Analysis

Laboratory Experiments in Trace Environmental Quantitative Analysis

Paul R. Loconto

CRC Press is an imprint of the
Taylor & Francis Group, an **informa** business

About the front cover: The author utilized the Multipurpose Sampler (Gerstel GmbH & Co.KG) based on the CTC PAL design atop a 6890/5973 GC-MS (Agilent Technologies) to meet the U.S. Centers for Disease Control and Prevention's (CDC's) demand for quantitating trace cyanide and trace VOCs in human blood and serum respectively.
Photo is courtesy of the Michigan Department of Community Health, Bureau of Laboratories, Lansing.

First edition published [2022]
by CRC Press
6000 Broken Sound Parkway NW, Suite 300, Boca Raton, FL 33487-2742

and by CRC Press
4 Park Square, Milton Park, Abingdon, Oxon, OX14 4RN

CRC Press is an imprint of Taylor & Francis Group, LLC

Library of Congress Cataloging–in–Publication Data

Names: Loconto, Paul R., 1947- author.
Title: Laboratory experiments in trace environmental quantitative analysis / Paul R. Loconto.
Description: First edition. | Boca Raton : CRC Press, 2022. | Includes bibliographical references and index.
Identifiers: LCCN 2021051243 (print) | LCCN 2021051244 (ebook) | ISBN 9781032197579 (hardback) | ISBN 9781032195629 (paperback) | ISBN 9781003260707 (ebook)
Subjects: LCSH: Environmental chemistry--Laboratory manuals. | Trace analysis--Laboratory manuals. | Analytical chemistry--Quantitative--Laboratory manuals.
Classification: LCC TD193 .L629 2022 (print) | LCC TD193 (ebook) | DDC 628.5--dc23/eng/20220124
LC record available at https://lccn.loc.gov/2021051243
LC ebook record available at https://lccn.loc.gov/202105124

ISBN: 9781032197579 (hbk)
ISBN: 9781032195629 (pbk)
ISBN: 9781003260707 (ebk)

DOI: 10.1201/9781003260707

Typeset in Times
by Deanta Global Publishing Services, Chennai, India

To

Priscilla Ann (Hamel) Loconto

My wife of nearly 50 years!

She has graciously supported her husband's career interests which eventually evolved into writing books while being a wonderful role model to our five daughters.

Contents

Preface

This laboratory manual was conceived out of a need to offer beginning graduate students who choose to specialize in environmental engineering at Michigan State University (MSU) a laboratory experience in learning to prepare samples as well as learning to quantify inorganic and organic chemical contaminants that might be present in surface water, groundwater, and contaminated soil. This laboratory-based initiative crosses the more traditional academic disciplines of environmental science and analytical chemistry. These experiments were conceived and implemented prior to the author writing a text and reference book on the same subject.[1]

University funds from the Department of Civil and Environmental Engineering within the College of Engineering at MSU were used to establish a teaching laboratory in the mid-1990s. This laboratory included, in addition to the necessary fume hoods and laboratory benches, four student workstations. Each student workstation included: one gas chromatograph; one high-performance liquid chromatograph; and one atomic absorption spectrophotometer. Several ultraviolet-visible spectrophotometers and several pH meters were made available for student use throughout the laboratory. In addition, an ion chromatograph and a gas chromatograph–mass spectrometer were made available for use by staff with possible use by students. This author had the opportunity to teach the course titled: ENE 807 (lecture) and ENE 808 (lab), shortly after the laboratory installations were completed.

The content of this laboratory manual is also appropriate for undergraduate students who major in chemistry with an emphasis on helping such students to develop laboratory aptitude and skill.

Educationally, we can begin to address environmental contamination, which inevitably leads to environmental pollution, by learning how to conduct *trace environmental quantitative analysis*. This initiative is consistent with the United Nations (UN) Goal #3: Good Health and Well Being; UN Goal #4: Quality Education; and UN Goal #6: Clean Water and Sanitation.

Shortly after the ENE teaching laboratory was established in the main engineering building on the MSU campus, a series of photos were taken. These photos show the author with his two assistants engaged with the analytical instrumentation at that time.

NOTE

1. Loconto P.R. *Trace Environmental Quantitative Analysis: Including Student-Tested Experiments*, 3rd ed. Boca Raton, FL: CRC Press, 2021.

About the Author

Paul R. Loconto holds a Ph.D. in analytical chemistry from the University of Massachusetts Lowell and an M.S. in physical organic chemistry from Indiana University, Bloomington. He has published 35 peer-reviewed papers in analytical chemistry and in chemical education. He has given over 40 talks and poster presentations at various workshops, meetings, and conferences.

After brief stints at the American Cyanamid Co. (Stamford, CT) and the Dow Chemical Co. (Midland, MI) and beginning in 1974, Dr. Loconto taught introductory, general, and organic chemistry at Dutchess Community College (Poughkeepsie, NY) for 12 years. He developed an environmental analysis laboratory course for the college's natural resource conservation program. He also expanded the college's capability in analytical instrumentation. He then joined NANCO Environmental Services (Wappingers Falls, NY) as R&D manager in 1986.

While at NANCO, he received a Phase 1 Small Business Innovation Research (SBIR) Grant from the US Environmental Protection Agency (US EPA) to conduct trace analytical method development to identify and quantify traces of environmental organic contaminants in water using reversed-phase solid-phase extraction (RP-SPE) techniques. RP-SPE was relatively new back then. The New York State Science and Technology Foundation matched the Phase 1 SBIR grant at that time. He also contributed to initiating a technical library while contributing to the laboratory's training initiatives.

He joined the Michigan Biotechnology Institute (Lansing) in 1990 where he led a small analytical research group in support of pilot-scale fermentation studies in agricultural biotechnology.

In 1992 he became the laboratory manager of the analytical laboratory in graduate research for environmental engineering at Michigan State University (East Lansing) where he conducted trace analytical method development for both the National Institute of Environmental Health Science (NIEHS) analytical core and the US EPA Hazardous Substance Research Center while coordinating the development of an instructional analytical laboratory for the graduate school.

In 2001, he joined the Michigan Department of Community Health Bureau of Laboratories (Lansing) as a Laboratory Scientist Specialist. Here, in addition to training new employees on how to use gas chromatographs, gas chromatograph–mass spectrometers, high-performance liquid chromatographs, and Fourier Transform infrared spectrophotometers as well as RP-SPE sample prep techniques, he taught co-workers how to satisfy QA/QC requirements. He also focused on developing novel analytical methods for biomonitoring while conducting trace organics and inorganics quantitative analysis in support of the Laboratory Response Network for the US Centers for Disease Control and Prevention (CDC).

He retired in late 2013 yet continues as a consultant, educator, and writer.

Photos

The following photos were taken of the author, Paul R. Loconto, Ph.D., and his two assistants, Yan Pan and Joseph Nguyen, during the mid-to-late 1990s. Dr. Loconto worked, at that time, as an analytical chemist while he was employed as a Laboratory Manager/Research Assistant in the Department of Civil and Environmental Engineering, College of Engineering, Michigan State University, East Lansing from 1992–2001.

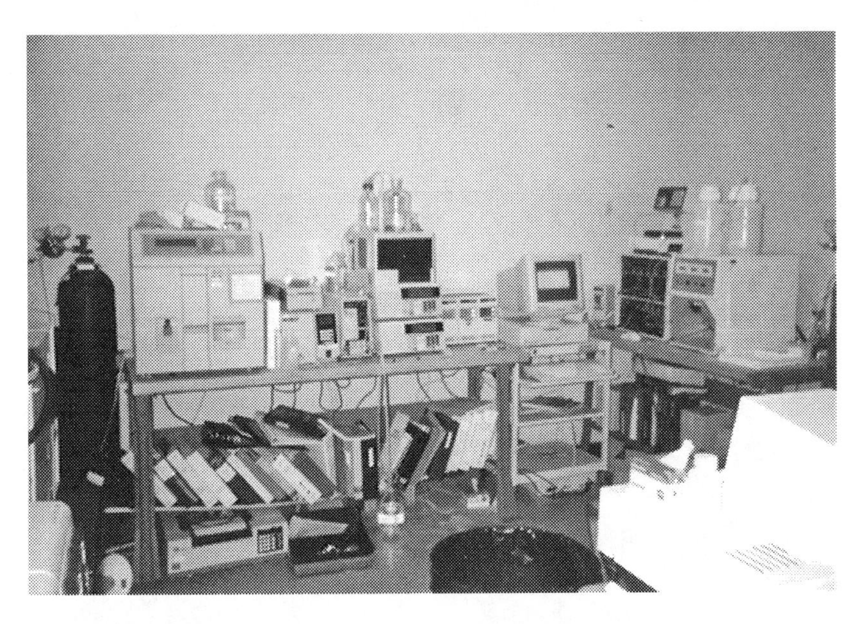

PHOTO #1 shows a section of a working analytical instrumentation laboratory located at the Research Complex, Department of Civil and Environmental Engineering, Michigan State University, East Lansing. From left to right: a capillary electrophoresis instrument, a high-performance liquid chromatograph to include UV-visible and fluorescence detectors, an ion chromatograph, and a light scattering detector all integrated into a single computer via an analog to digital (A to D) interface. In the foreground, a stand-alone UV-vis absorbance spectrophotometer is shown.

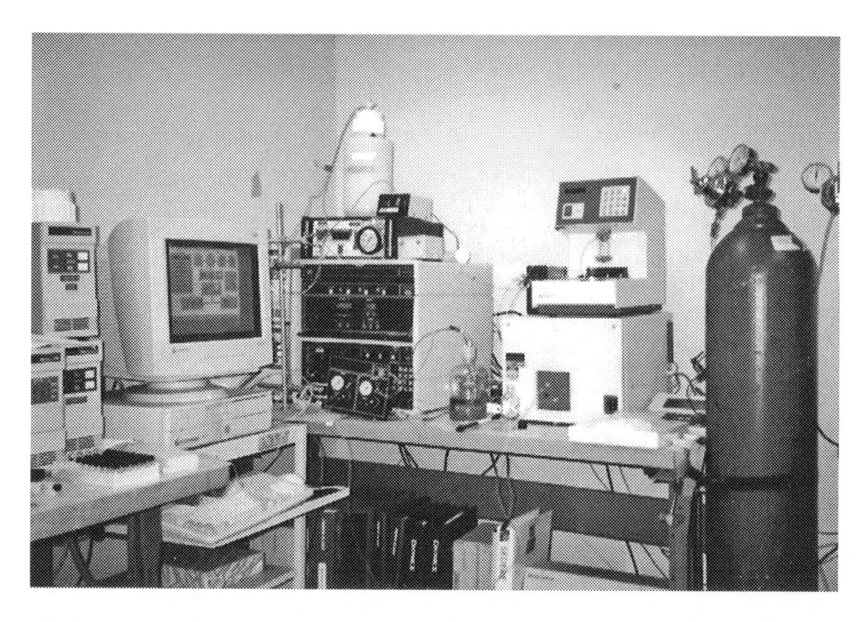

PHOTO #2 shows a close-up at the same location as shown in Photo #1 with an emphasis on the right corner of the laboratory. The light scattering detector has been replaced by an auto-sampler and liquid chromatographic column atop a UV-vis flow-through HPLC photometric detector (RP-HPLC-UV).

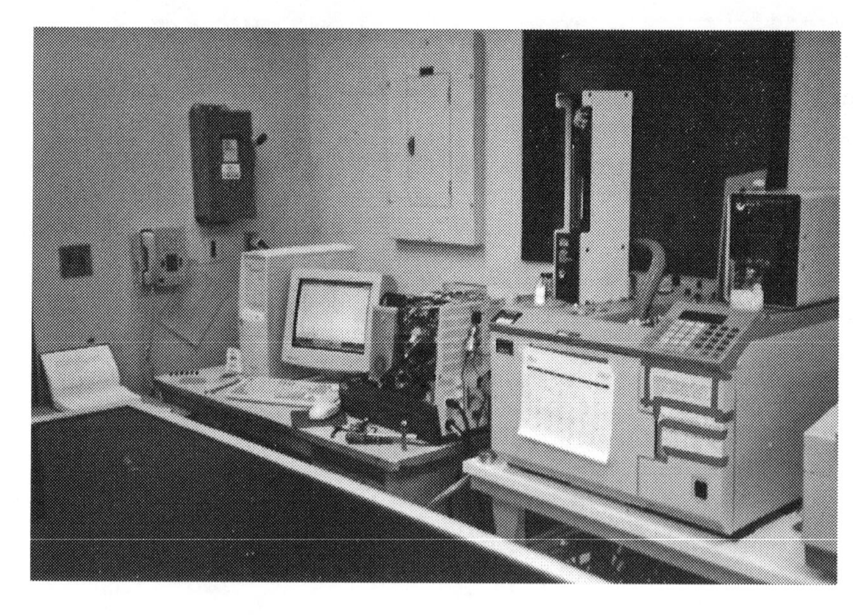

PHOTO #3 at the same location shows from right to left, a gas chromatograph interfaced to an ion-trap mass spectrometer (GC-MS) with the cover removed which is interfaced via an analog to digital converter to a computer.

PHOTO #4 depicts the same instrumentation as Photo #3 shown from a different perspective.

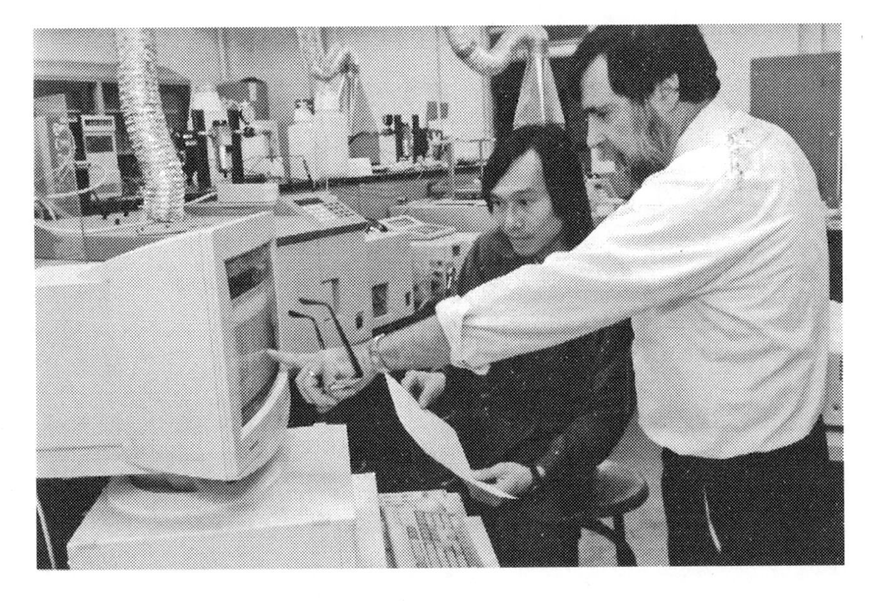

PHOTOS #5 is located at the environmental engineering teaching analytical laboratory (ENE Lab), in the main engineering building, and shows the author with Joseph Nguyen located at one of the four student workstations.

PHOTO #6 shows Joseph Nguyen operating a graphite furnace atomic absorption spectrophotometer at one of the four student workstations located in the ENE Lab, main engineering building.

PHOTO #7 shows Yan Pan, Joseph Nguyen, and the author located at one of the four student workstations in the ENE Lab, main engineering building. Just behind the author is a Perkin-Elmer inductively coupled plasma atomic emission spectrometer (ICP-AES).

PHOTO #8 shows Yan Pan pipetting samples at the Research Complex.

PHOTO #9 shows Yan Pan and the author in front of a purge and trap concentrator interfaced to an Agilent Technologies gas chromatograph–mass spectrometer (GC-MS) located at the Research Complex.

PHOTO #10 shows Yan Pan and the author at a Perkin-Elmer graphite furnace atomic absorption spectrophotometer located at the Research Complex.

PHOTO #11 shows Yan Pan, Joseph Nguyen, and the author in front of a Perkin-Elmer carbon, hydrogen, nitrogen (CHN) analyzer, located in the civil and environmental engineering analysis laboratory (CEE Lab), main engineering building.

1 Introduction

Theory guides, experiment decides.

—I.M. Kolthoff

This chapter introduces a series of laboratory experiments that attempt to show some examples of how to conduct trace environmental quantitative analysis (TEQA). These experiments were originally incorporated in the last chapter of the author's text, *Trace Environmental Quantitative Analysis*, now in its 3rd edition. These experiments address only the *enviro-chemical* aspects of TEQA. The impetus for writing these experiments was in support of a graduate-level course titled Environmental Analytical Chemistry Laboratory. This course began in the mid-1990s, and the instruction followed the installation of a unique teaching laboratory coordinated by the author while employed as a laboratory manager/research assistant for the Hazardous Substance Research Center. This federally funded center was a focus of research and development in the Department of Civil and Environmental Engineering at Michigan State University, East Lansing.

There are several options that an instructor can use to design a laboratory program that gives M.S. and Ph.D. graduate students numerous opportunities to measure environmentally significant chemical analytes. It is this author's opinion that it doesn't really matter which analytes are to be quantitated as long as an *appropriate mix of sample prep and instrumental techniques* is introduced to students who will become environmental engineers, water treatment plant operators, and environmental analytical chemists, among other occupational endeavors. A suggested laboratory schedule that was successfully implemented earlier is introduced in the following sections.

1.1 WHAT MIGHT A TYPICAL LABORATORY SCHEDULE LOOK LIKE?

The laboratory program once implemented by the author for a graduate-level course in TEQA is shown in the table that follows. These are indeed *student-tested experiments*! Beneath the title of each experiment is a statement of learning objectives and outcomes that the student should be able to achieve. The degree to which the instructor makes the course *more or less rigorous* is determined by the curriculum objectives. An experimental course in TEQA can consist of a series of experiments with everything set up beforehand for the student. Alternatively, the same experiments can be undertaken whereby the students do all the setting up themselves! Some compromise between these two extremes might be in the best interest of both the student and laboratory staff alike! A series of actual student experiments given as *individual handouts* follows the laboratory course annotated syllabus and statement of learning objectives as shown for each laboratory session shown here.

DOI: 10.1201/9781003260707-1

Experiment #	Description
1st half or full semester	Orientation to laboratory; discussion of outcomes and what is expected; definition of learning goals and student assignment to workstations; safety requirements; how to weigh properly, guidelines for safe and environmentally sound disposal of laboratory-generated chemical waste. *Descriptive introductory information. Instructors should also emphasize the safety aspects of working with chemicals, laboratory equipment, and sophisticated and expensive analytical instrumentation.*
1	Introduction to pH measurement: estimating the degree of purity of snow; learning to measure soil pH; introduction to ion chromatography *and/or* introduction to visible spectrophotometry *and/or* determination of the Fe(III)/Fe(II) concentration ratio in groundwater *and/or* determination of phosphate ion, PO3–4 in eutrophicated surface water. *Quantitative analysis; emphasis on standards preparation techniques; statistical treatment of data; environmental sampling techniques; learning to operate a pH meter; learning to operate the UV-vis spectrophotometer; learning to operate a flame atomic absorption (AA) spectrophotometer; no write-up required.*
2	Determination of anionic surfactants in an industrial wastewater effluent by mini-liquid–liquid (mini-LLE) extraction using ion pairing with methylene blue. *Quantitative analysis; emphasis on sample preparation, forming an ion pair, unknown sample analysis; learning to use and to calibrate a pH meter; measuring visible absorbance on a spectrophotometer; write-up required.*
3	Comparison of ultraviolet and infrared absorption spectra of chemically similar organic compounds *and/or* determination of oil and grease and of total petroleum hydrocarbons (TPHs) via reversed-phase solid-phase extraction (RP-SPE) techniques using quantitative Fourier-Transform infrared (FTIR) spectroscopy.* *Qualitative analysis; background for understanding UV-vis and infrared absorption spectroscopy; learning to use and interpret data from molecular spectroscopic instrumentation; sampling techniques; write-up required.*
4	Determination of the degree of hardness in groundwater using flame atomic absorption (FLAA) spectroscopy: measuring Ca, Mg, and Fe. *Quantitative analysis; learning to operate a FLAA spectrophotometer; calibration using external standard mode; spiked recovery; no write-up required.*
5	Determination of lead (Pb) in drinking water using graphite furnace atomic absorption spectroscopy (GFAA). *Quantitative analysis; learning to operate a GFAA spectrophotometer; learning to use the WinLab® software for GFAA spectrophotometry; calibration based on standard addition; no write-up required.*
6	Comparison of soil types via a quantitative determination of the chromium content using visible spectrophotometry and FLAA spectrophotometry or inductively coupled plasma-atomic emission spectrometry. *Quantitative analysis; use of two instrumental methods to determine the Cr (III) and Cr (VI) oxidation states; digestion techniques applied to soils; write-up required.*

(*Continued*)

Experiment #	Description
2nd half or full semester	
7	An introduction to data acquisition and instrument control using Turbochrom®; introduction to high-performance liquid chromatography (HPLC); evaluating those experimental parameters that influence HPLC instrument performance. *Qualitative analysis; emphasis on learning to operate the HPLC instrument and the Turbochrom® computer software; no write-up required; answer questions in lab notebook.*
8	Identifying the ubiquitous phthalate esters in the environment using HPLC with photodiode array detection (PDA); confirmation using gas chromatography-mass spectrometry (GC-MS). *Qualitative analysis; interpretation of chromatograms, UV absorption spectra, mass spectra; experience with GC-MS; write-up required.*
9	An introduction to gas chromatography: evaluating experimental parameters that affect gas chromatographic performance. *Qualitative analysis; emphasis on learning to operate the GC; measurement of split ratio; no write-up required; answer questions in lab notebook.*
10	Screening for the presence of BTEX in gasoline-tainted water using mini-LLE and gas chromatography (GC-FID) *and/or* screening for THMs in chlorine-disinfected drinking water using static HS gas chromatography (GC-ECD). *Semi-quantitative analysis; unknown sample analysis; statistical treatment of data; write-up required.*
11	Determination of priority pollutant volatile organic compounds (VOCs) in gasoline-contaminated groundwater using static headspace (HS) and solid-phase microextraction headspace (SPME-HS) and gas chromatography. *Quantitative analysis, learning to use HS and SPME-HS sample prep techniques; learning to inject an HS syringe into a gas chromatograph; write-up required.*
12	Determination of the concentration of the herbicide residue trifluralin in soil from lawn treatment by gas chromatography using reversed-phase solid-phase extraction (RP-SPE) methods. *Quantitative analysis; calibration based on internal standard mode; unknown sample analysis; statistical treatment of data; write-up required.*
13	Determination of priority pollutant semivolatile (SVOC) organochlorine pesticides in contaminated groundwater: comparison of two sample preparation methods—mini-LLE vs. RP-SPE techniques.* *Quantitative analysis; emphasis on sample preparation, unknown sample analysis; calibration based on internal mode; statistical treatment of data; write-up required.*
14	Determination of selected priority pollutant polycyclic aromatic hydrocarbons in oil-contaminated soil using LLE-RP-HPLC-PDA. *Quantitative analysis; sample preparation; write-up required.*

* Projects are considered extra credit and thus not required. Students must make arrangements with the laboratory instructor in order to perform these experiments.

This is a very ambitious laboratory course! Instructors may wish to run the laboratory course for two semesters in one academic calendar year. It is, however, a graduate-level laboratory course of instruction. To effectively educate graduate students while delivering the course content requires a dedicated support staff, a committed faculty, sufficient laboratory glassware and accessories, and expensive analytical instrumentation, including the interface of each instrument to a personal computer that operates chromatography or spectroscopy software. Each laboratory session requires a minimum of four hours and a maximum of eight hours. Students must be taught not only how to *prepare environmental samples for trace analysis*, but also how to *operate and interpret the data* generated from sophisticated analytical instruments. The intensity of the lab activities starts with an initial and less rigorous laboratory session/experiment with rigor increasing as each laboratory session/experiment unfolds.

1.2 HOW IS THE INSTRUCTIONAL LABORATORY CONFIGURED?

When the laboratory experiments that follow were developed, the author had just completed coordinating the installation of an instructional laboratory that included four workstations. In addition, one ICP-AES, one FTIR spectrophotometer, one ion chromatograph, and several UV-vis spectrophotometers were available for student use.

Each workstation consisted of:

1 One Autosystem® (PerkinElmer Instruments) gas chromatograph incorporating dual capillary columns (one for VOCs and one for SVOCs) and dual detectors (FID and ECD).
2 One HPLC (PerkinElmer Instruments) that included a 200 Series® LC binary pump, a manual injector (Rheodyne®), a reversed-phase column and guard column, and a LC250® photodiode array (PDA) ultraviolet absorption detector.
3 One Model 3110® (PerkinElmer Instruments) atomic absorption spectrophotometer with flame and graphite furnace capability with deuterium background correction.
4 One personal computer (PC) that enabled all three instruments above to be interfaced. For GC and HPLC, Turbochrom® (PE Nelson) Chromatography Processing Software (now called Total Chrom; PerkinElmer Instruments) was used for the data acquisition via the 600 LINK® (PE Nelson) interface that was external to the PC. For AA, WinLab® (PerkinElmer Instruments) software was used via an interface board that was installed into the PC console.
5 A UV-VIS spectrophotometer Genesys 5® (Spectronic Instruments) was used. If another spectrophotometer is used, an infrared phototube is necessary to quantitate in that experiment. Multiple spectrophotometers would enhance instruction.

Each student workstation was configured as shown in the schematic below where *A/D* refers to an analog-to-digital interface:

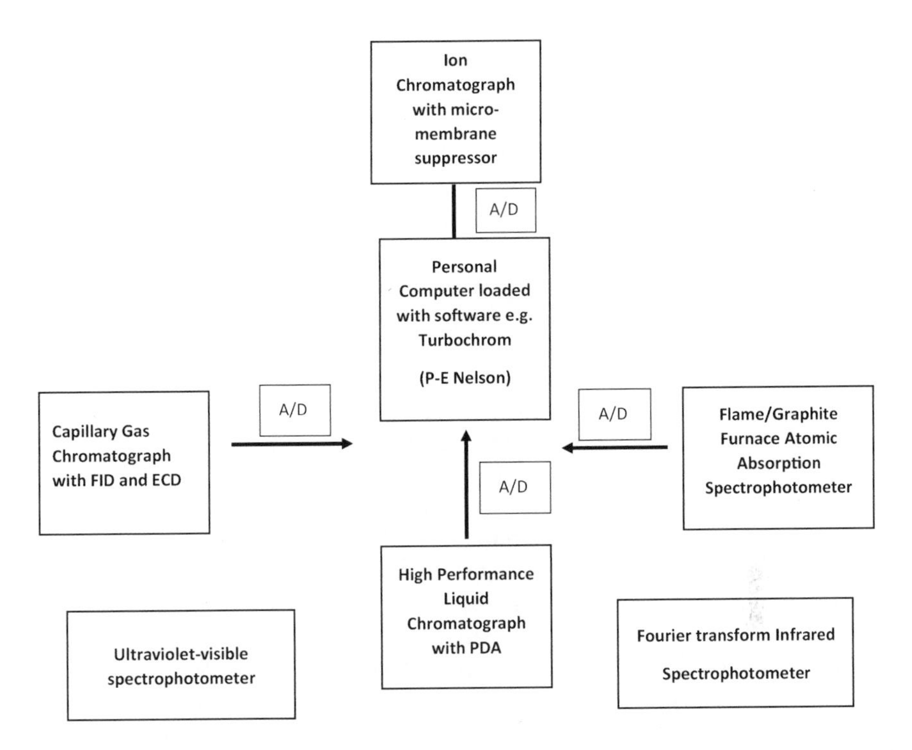

Analytical instrument configuration for the teaching laboratory: instruments connected with arrows can simultaneously send analytical data to the central computer via A/D converters.

This configuration was the one used at Michigan State University in the College of Engineering and within the Department of Civil and Environmental Engineering at that time. Individual college departments will undoubtedly have their own unique laboratory configurations.

In addition, a Model 2000® (Dionex) ion chromatograph interfaced to the PC via a 900® interface (PE Nelson) and a Model 1600® FTIR Spectrophotometer (PerkinElmer Instruments) are available for all students to use in the instructional laboratory. Individual university and college departments will undoubtedly have their own unique laboratory configurations that include *instrumentation from other manufacturers*. In order to carry out all of the experiments introduced in this chapter, instructional laboratories must have, at a minimum, the following analytical instruments: GC-FID/ECD, HPLC-UV or HPLC-PDA, FlAA and GFAA, IC, and a UV-vis spectrophotometer (stand-alone). Access to a GC-MS is also required in one experiment. Accessories for sample preparation, as listed in each of the subsequent experiments, are also needed.

Each experiment that follows was written to be as independent of the others in the collection as possible. Accompanying each experiment is a short list of references. Safety tips appear in each experiment as poignant reminders to students and instructors alike of the *perils associated with laboratory work*. Instructors can pick and choose to use a given experiment as written here or modify it to fit their unique laboratory situation. Students are encouraged to learn and use, in addition to mastering the software associated with each instrument. The software tool Excel® is used to facilitate data analysis and interpretation. Most student experiments shown here end with a list of books and/or literature references cited as *suggested readings* that relate to that specific experiment while serving to broaden the scope of the given topic. As part of an introductory exercise, students should be introduced to the proper use of a contemporary analytical balance.

1.3 HOW TO WEIGH THE RIGHT WAY

The analytical balance has been and continues to be the simplest yet most profound instrument in the analytical laboratory. The precision and accuracy of the most sophisticated analytical instrument are limited by the precision and accuracy inherent in the weight of the neat form of a certified reference standard. Today's single pan electronically digitized balances are a *far cry* from the analog double pan analytical balance that required the very careful adding of various-sized calibration weights placed on the right-hand pan to balance the weight of samples placed on the left-hand pan. The procedure shown next was adapted from I. Ciesniewski, *R&D Magazine* 45:30 (2003).

The contemporary electronic analytical balance must be properly maintained, cleaned frequently, calibrated with standard reference weights, and used properly. Some helpful hints on developing proper weighing skills are given here:

- Before weighing, ensure that your balance is correctly leveled.
- Periodically check, clean, and calibrate the balance.
- Zero out or tare the balance prior to weighing.
- Minimize the use of hands to place tare weights or samples in the weigh chamber. Use appropriately sized and shaped tweezers or tongs to handle weighing vessels.
- Weigh to the same side each and every time. This will maximize repeatability.
- When placing items to be weighed onto the weigh pan, open only the draft shield door on the side on which you are weighing: e.g. if you are right-handed, open the right door.
- Understand how the balance indicates a stable weight, i.e. gives a weight that can be safely trusted. All electronic balances give a visual indication of weight stability.
- Zero out or tare weight as carefully as done for the sample. Using tweezers, place the tare weight onto the weigh pan, close all doors, press the tare button, and wait for the balance to give a stable zero.
- Introduce the sample into the weighing vessel using a long spatula, spoon, scoop, or tweezers as appropriate.

- Be aware of how the balance is affected by the working environment. Modern, busy laboratories are not ideal places for the four- or five-decimal-place balances that we need to put in them. It may be difficult to stabilize. It may take as long as 60 seconds for the balance to reach stability.
- Aim for the same location on the balance weigh pan; try to aim for the center of the pan each time.
- After finishing weighing, check that the weigh chamber is clean and free of any spills. This is not just a courtesy to others or conformance to regulations; it is helping to keep your balance working accurately by eliminating unwanted ingression that could damage the internal mechanics of the balance.

2 An Introduction to pH Measurement

Estimating the Degree of Purity of Snow, Measuring Soil pH: Introduction to Ion Chromatography

2.1 BACKGROUND AND SUMMARY OF METHOD

This laboratory will focus on the operational aspects of pH measurement. It is appropriate that we start this course with pH because this parameter is so fundamental to the physical-chemical phenomenon that occurs in aqueous solutions. The pH of a solution that contains a weak acid determines the degree of ionization of that weak acid. Of environmental importance is an understanding of the acidic properties of carbon dioxide. The extent to which gaseous CO_2 dissolves in water and equilibrates is governed by the Henry's law constant for CO_2. We are all familiar with the carbonation of beverages. The equilibrium is

$$CO_{2(g)} \leftrightarrow CO_{2(aq)}$$

When CO_2 is dissolved in water, a small fraction of the dissolved gas exists as *carbonic acid*, an unstable substance whose theoretical chemical formula is written as H_2CO_3. The acid–base character of CO_2 is described more accurately by the following reactions and equilibrium constants:

$$CO_{2(aq)} + H_2O \leftrightarrow H^+_{(aq)} + HCO^-_{3(aq)}$$

$$K_{a1} = \frac{\left[H^+\right]\left[HCO_3^-\right]}{\left[CO_2\right]} = 4.45 \times 10^{-7}$$

$$pK_{a1} = 6.35 \tag{2.1}$$

DOI: 10.1201/9781003260707-2

$$HCO^{-}_{3(aq)} \leftrightarrow H^{+}_{(aq)} + CO^{2-}_{3(aq)}$$

$$pK_{a2} = 10.33 \qquad (2.2)$$

For example, the molar concentration of CO_2 in water saturated with this gas at a pressure of 1 atm at 25° C is 3.27 × $10^{-2}M$. The pH, using equation (2.1), can be calculated as 3.92. Environmental water samples generally have less than the saturated molarity value and yield a pH of approximately 5.9. Thus, the difference between *ultrapure water*, with a theoretical pH of 7, and water exposed to the atmosphere, with a measurable pH of 5.9, is attributed to dissolved CO_2 and its formation of *carbonic acid*, which in turn dissociates to *hydronium*, *bicarbonate*, and *carbonate* ions. The degree to which carbon dioxide dissolved in water remains in its molecular form or exists as either *bicarbonate* or *carbonate* depends upon the pH of water. The fraction, α, of each of these chemical species of the total is mathematically defined as

$$\alpha_{CO_2} = \frac{[CO_2]}{[CO_2]+[HCO_3^-]+[CO_3^{2-}]} \qquad (2.3)$$

$$\alpha_{HCO_3^-} = \frac{[HCO_3^-]}{[CO_2]+[HCO_3^-]+[CO_3^{2-}]} \qquad (2.4)$$

$$\alpha_{CO_3^{2-}} = \frac{[CO_3^{2-}]}{[CO_2]+[HCO_3^-]+[CO_3^{2-}]} \qquad (2.5)$$

Equations (2.3) to (2.5) can be reworked in terms of acid dissociation constants and hydronium ion concentrations to yield

$$\alpha_{CO_2} = \frac{[H^+]^2}{[H^+]^2 + K_{a1}[H^+] + K_{a1}K_{a2}} \qquad (2.6)$$

$$\alpha_{HCO_3^{2-}} = \frac{K_{a1}[H^+]}{[H^+]^2 + K_{a1}[H^+] + K_{a1}K_{a2}} \qquad (2.7)$$

$$\alpha_{CO_3^{2-}} = \frac{K_{a1}K_{a2}}{[H^+]^2 + K_{a1}[H^+] + K_{a1}K_{a2}} \qquad (2.8)$$

Equations (2.6) to (2.8) can be used to construct what are called *distribution of species diagrams*, which are plots of the fraction of each species as a function of solution pH. For the $CO_2 / HCO_3^- / CO_3^{2-}$ aqueous solution, the distribution of species diagram is shown below:

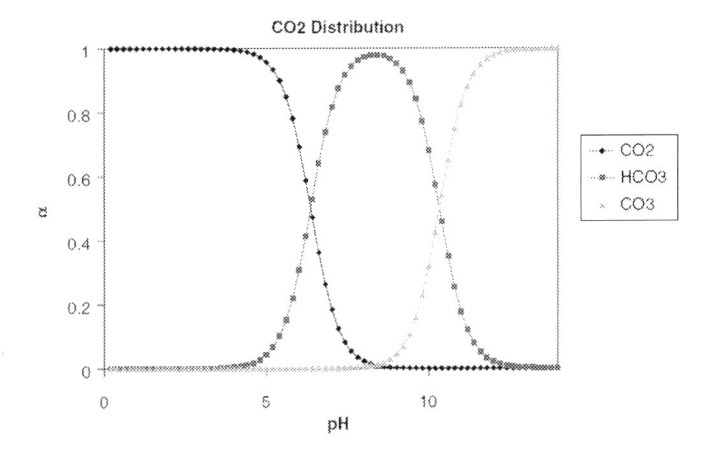

The *amount of acidic or alkaline solutes* present in a given volume of groundwater, surface water, or wastewater is determined by conventional acid–base titration. Refer to *Standard Methods for the Examination of Water and Wastewater* for specific procedures. Titrations are performed in these methods to a specified pH value. Be sure to distinguish among the concepts of acidity, alkalinity, and pH when considering the nature of environmental samples. The pH will need to be measured and adjusted prior to conducting the ion-pair liquid–liquid extraction for determining methylene blue active surfactants.

The pH was originally measured by judicious choice of acid–base indicator dyes and eventually over the years gave way to potentiometric methods due to the *inherent limitations of color*. A dye serves no useful purpose when the wastewater sample is brownish in color. Advances in both the instrument and the glass electrode have taken the measurement of pH very far since the early days when, in 1935, Arnold Beckman was first asked to measure the pH of a lemon!

Schematic drawings of the glass electrode and associated electronics are shown below:

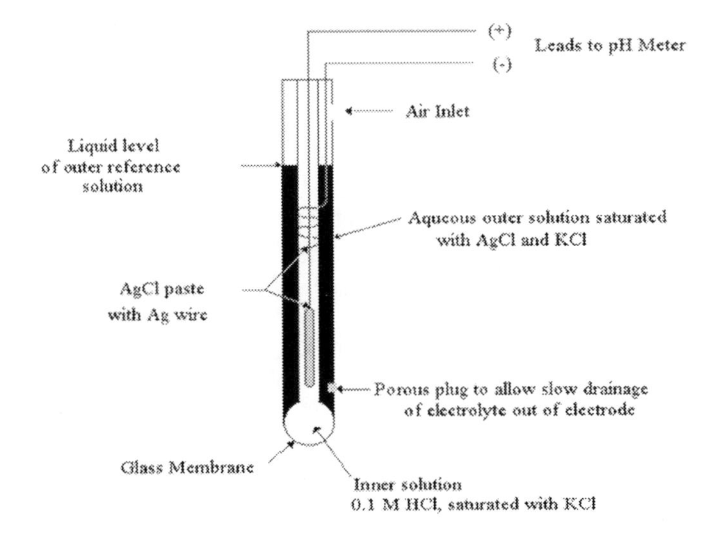

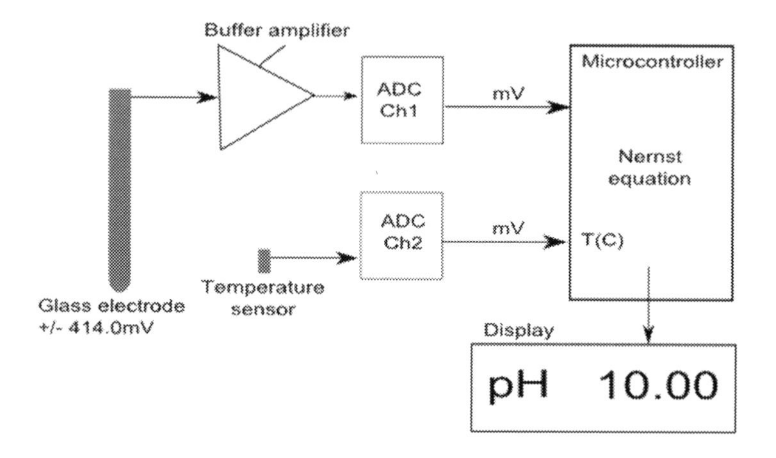

2.2 EXPERIMENT

2.2.1 Glassware Needed per Student

One 250 or 500 mL beaker to sample snow.
One 50 mL beaker.
One stirring rod.

2.2.2 Chemical Reagents/pH Meter Needed per Student Workstation

Each pH measurement station should consist of the following:

1 Three buffer solutions of pH 4, 7, and 10, respectively.
2 One squeeze bottle containing distilled deionized water (DDI).
3 One waste beaker.
4 Training guide for operating the Orion SA 720 pH/ISE® Direct Readout Meter.

2.2.3 Ion Chromatograph

A 2000® Dionex Ion Chromatograph will be set up by the staff for use in this experiment. As of this writing, what was then called the Dionex Corporation is now part of Thermo Fisher Scientific. Students are referred to the supplement titled *How to Set Up and Operate an Ion Chromatograph* which is found at the end of this laboratory manual and includes specifics on how to operate the instrument and how to prepare calibration standards.

A schematic drawing of a typical ion chromatograph is shown below:

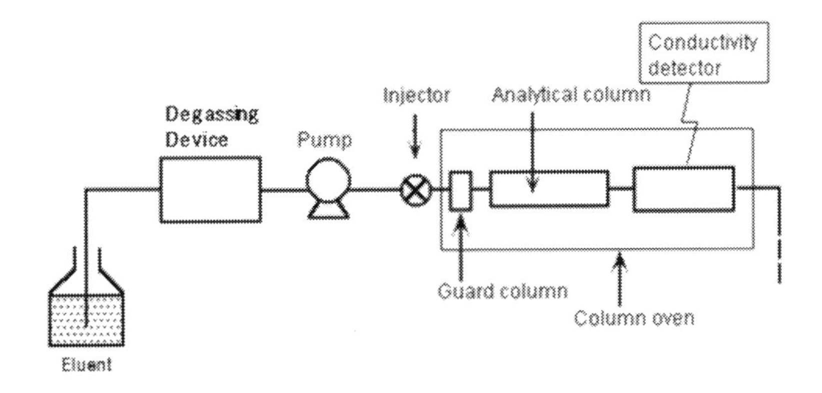

2.2.4 Procedure

Use the guide located at each of four pH measurement stations and familiarize yourself with the operation of the pH meter. Each student should calibrate the meter using the pH 4 and 7 buffer solutions that are available by implementing the auto-calibration. An ATC probe will not be available; therefore, samples and standards should be at the same temperature. Use a stirring rod to continuously stir the solution in the 50 mL beaker while a pH measurement is being taken.

Obtain at least three samples of snow. One sample should be obtained close to a walkway or roadway and appear visibly dirty. One sample should be obtained far away from pollutant sources and appear visibly clean. Use a relatively large beaker to collect snow and allow the snow to melt. *Record pH values in your notebook*. Draw conclusions about your observations and write these in your laboratory notebook.

Obtain one or more soil samples from a hazardous waste site. Alternatively, your instructor may have a series of fortified soils available in the laboratory. An illustration of how a series of laboratory-fortified acidic, neutral, and alkaline soils can be prepared and given to students as sample unknowns is shown in the following table. Students should review how to implement EPA Method 9045D from the EPA's SW-846 solid waste analysis methods. Today, students can quickly access most EPA Methods via a "Google name search." Assume that the unknown soil is *non-calcareous*. Notice the use of flowcharts in helping to understand the procedural aspects of the method. Repeat the pH measurement for the four other samples and *record your results* in your laboratory notebook. The chemical nature of the contaminated soil samples will be revealed to you after you have completed your pH measurements. Rationalize the observed pH value for each soil based on knowledge of the chemical used to contaminate the soil. Be sure to jot down your comments in your laboratory notebook.

Sample #	Soil Type	Chemical Added
1	Tappan B	Salicylic acid
2	Tappan B	Sodium carbonate
3	Tappan B	None
4	Hudson River sediment, possibly contaminated with PCBs	None
5	Unknown sandy soil	Citric acid
6	Unknown sandy soil	Potassium hydrogen phthalate
7	Unknown sandy soil	None
8	Unknown sandy soil	p-Nitrophenol
9	Unknown sandy soil	Tetrabutylammonium hydroxide

Molecular structures for the organic compounds used to adulterate the soil with the purpose of simulating a chemically polluted soil sample are given below:

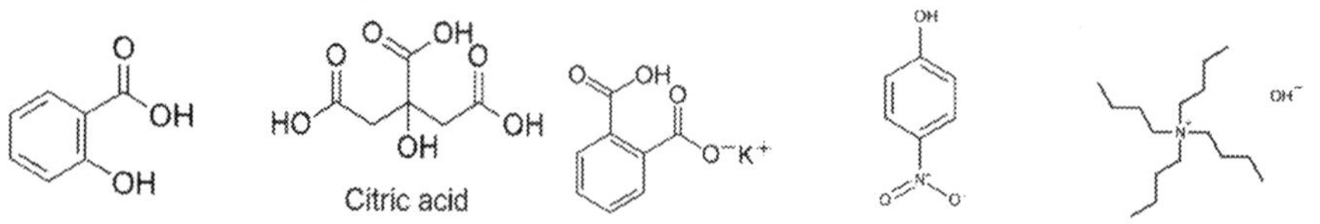

Salicyclic acid/citric acid/potassium hydrogen phthalate/p-nitrophenol/tetrabutyl ammonium hydroxide

As an additional feature to this experiment, the environmental engineering staff will have a Dionex 2000® Ion Chromatograph in operation. The staff will engage students in the operation of the ion chromatograph. This instrument will separate and detect inorganic anions such as chloride, nitrate, and sulfate. Are these chemical species present in the various snow samples? In what manner do inorganic anions contribute to snow acidity?

You have generated hazardous waste from your soil pH measurements. Dispose of it properly in accordance with the Office of Radiation, Chemical, and Biological Safety (ORCBS) guidelines.

The ORCBS provides institutional guidance and safety compliance to all research and teaching laboratories at Michigan State University.

2.3 SUGGESTED READING

To develop this experiment, the author consulted the following resources:

Many of the more lucid presentations on pH are found in general chemistry texts, whereas instrumental concepts of pH measurement are found in most texts on analytical chemistry.

The most definitive text at an advanced level is Bates, R. *Determination of pH*. New York: Wiley, 1964. This book is found in most university chemistry libraries.

A useful guide to the measurement of pH in soil is found in EPA Method 9045D. *Test Methods for Evaluating Solid Waste: Physical/Chemical Methods*. SW-846. Revision 4, November 2004.

Weiss J. *Handbook of Ion Chromatography*. E.L. Johnson Ed. Sunnyvale, CA: Dionex Corporation, 1986. There was no better treatment of the subject at that time.

3 Introduction to the Visible Spectrophotometer

3.1 BACKGROUND AND SUMMARY OF METHOD

The simple visible spectrophotometer has been an important instrument in trace environmental quantitative analysis for many years. Colorless environmental contaminants must be chemically converted to a species that appears colored to the eye. The colored species will then exhibit regions of the visible electromagnetic spectrum where it will absorb photons. This absorption can be related to the Beer–Lambert or Beer–Bouguer law of spectrophotometry according to:

$$A = \varepsilon bc$$

where A is the absorbance measured in absorbance units (AUs), and in turn is related to the logarithm of the ratio of incident to transmitted intensity; ε is the molar absorptivity, a unique property of a chemical substance, viewed as a constant in the equation; b is the path length in either millimeters or centimeters, assumed to be a constant (see cuvette matching below); and c is the concentration in either molarity or weight/unit volume, if known. This forms the basis for the calibration of *UV-visible and atomic absorption spectrophotometers.*

Recall that *the color of a solution is the complement of the color of light that it absorbs*. For example, the red cobalt chloride solution you will be using to conduct the cuvette matching (see the following) actually absorbs the complement to red light, which is green light at a wavelength of 510 nm. The term *spectrophotometer* succinctly and completely describes the instrument. *Spectro* refers to the visible spectrum and the ability of the instrument to select a wavelength or, more accurately, a range of wavelengths. *Photo* refers to light, and *meter* implies a measurement process. Some instruments lack the capability of selecting a wavelength and should be called photometers. A good question to ask when using a spectrophotometer is: what is the precision in nanometers when I set the wavelength? The answer should be as follows: it depends on the *effective bandpass of the monochromator*. The effective bandpass is a measure of the band of wavelengths allowed to pass through the spectrophotometer for a given width of the exit slit of the monochromator. The bandpass is given by the product of the slit width and the reciprocal linear dispersion. The

DOI: 10.1201/9781003260707-3

theory of light dispersion on gratings and prisms and subsequent effects on monochromator resolution has been well developed.

The major limitation of the use of visible spectrophotometric or colorimetric methods is the fact that the analyte must absorb within the visible domain of the electromagnetic spectrum. *Many analytes of environmental interest are colorless!* Some analytes can be chemically converted to a colored product by a direct chemical change, formation of a metal chelate, or formation of an ion pair. An example of a chemical conversion of the analyte of interest is the colorimetric determination of nitrite–nitrogen via reaction of the nitrite ion with sulfanilic acid in acidic media to form the diazonium salt, with subsequent reaction with chromotropic acid to form a highly colored azo dye as shown:

$$NO_2^- + HO_3S-C_6H_4-NH_2 + 2H^+ \longrightarrow HO_3S-C_6H_4-N\equiv N^+ + 2H_2O$$

Nitrite Sulfanilic acid Diazonium salt

$$HO_3S-C_6H_4-N\equiv N^- + \text{(OH)}_2C_{10}H_4(SO_3H)_2 \longrightarrow \text{(OH)}_2C_{10}H_3(SO_3H)_2-N=N-C_6H_4-SO_3H + H^-$$

Diazonium salt Chromotropic acid Red-orange-colored complex

An example of metal chelation can be found if one refers to *Standard Methods for the Examination of Water and Wastewater* for the recommended methods for determining the toxic metal cadmium. An atomic absorption method is listed along with a dithizone method. Cd^{2+} combines with dithizone (HDz) in aqueous media to form the neutral-colored molecule $Cd(Dz)_2$. This neutral molecule can be subsequently extracted into a nonpolar solvent, thus providing an increase in the method detection limit (MDL). The extent to which a given metal ion can be chelated with HDz and subsequently extracted is defined as the distribution ratio, D_M, for a given metal ion and depends on the formation constant of the complex, the overall extraction constant, the initial concentration of HDz in the organic phase, the fraction of the total metal concentration present as the divalent cation, and, most importantly, the pH of the aqueous solution. For the equilibrium

$$\mathrm{M}^{n+}_{(\mathrm{aq})} + n\mathrm{HDz}_o \leftrightarrow \mathrm{MDz}_{n(o)} + nH^+_{(\mathrm{aq})}$$

the distribution ratio is expressed mathematically as

$$D_M = \frac{\beta_n K_{\mathrm{ex}} \left[\mathrm{HDz}\right]_o^n}{\left[\mathrm{H}^+\right]^n} \alpha_M$$

An example of a chelated ion association pair that when formed is subsequently extracted into a nonpolar solvent is that of the iron(II)-*o*-phenanthroline cation, which efficiently partitions into chloroform, provided a large counterion such as perchlorate is present in the chloroform.

3.2 EXPERIMENT

A possible major source of error in spectrophotometric measurement occurs when the path length differs from one sample to the next (i.e. the value for b [see earlier discussion] is different). Hence, two solutions having identical concentrations of a colored analyte (i.e. both a and c are equal) could exhibit different absorbances due to differences in path length b. To minimize this source of systematic error, *a cuvette matching exercise is introduced*. Cuvette tubes are matched by placing a solution of intermediate absorbance in each and comparing absorbance readings. One tube is chosen arbitrarily as a reference, and others are selected that have the same reading to within 1%. Tubes should always be matched in a separate operation before any spectrophotometric experiments are begun.

3.2.1 Glassware Needed per Student or Group

Seven to ten 13 × 100 nm glass test tubes.

3.2.2 Chemical Reagents Needed per Student or Group

Ten milliliters of a 2% cobalt chloride solution; dissolve 2 g of $CoCl_2{\bullet}6H_2O$ in approximately 100 mL of 0.3 *M* HCl. This solution is recommended because it is stable, has a broad absorption band at about the center of the visible region, and transmits about 50% in a 1 cm cell. Each group should prepare this solution and share it among members.

3.2.3 Miscellaneous Item Needed per Student or Group

One piece of chalk that nicely fits into a 13 × 100 mm test tube.

3.2.4 Spectrophotometer

Any stand-alone UV-vis absorption spectrophotometer can be used to perform this experiment. Students will be introduced to the *Spectronic 20*, commonly called a *Spec 20*. This instrument is of historical as well as educational value since it is the *most common spectrophotometer ever to be found in undergraduate college chemistry labs*. Shown in the following figure are schematic drawings of the original analog electronic instrument usually placed right on a lab benchtop followed by an optical schematic of the light path inside this "black box."

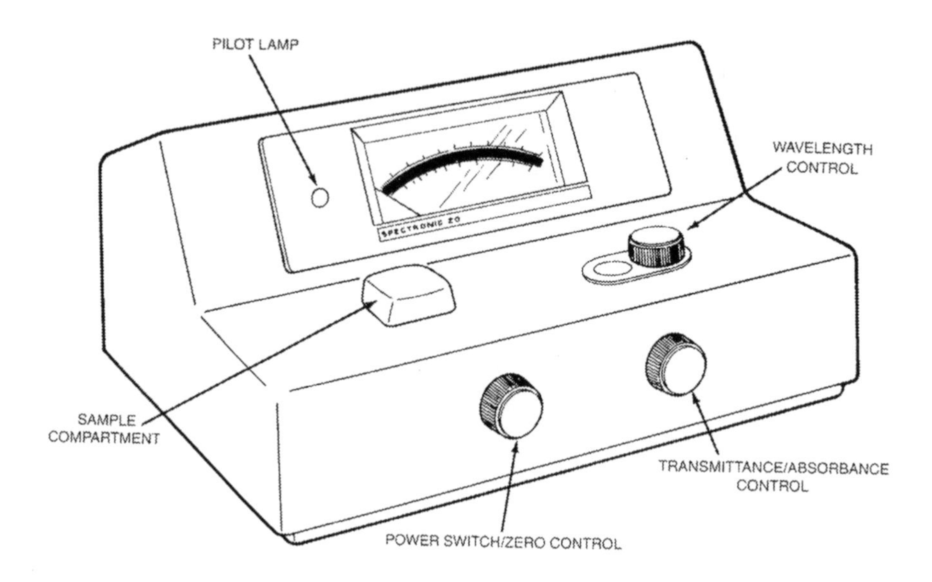

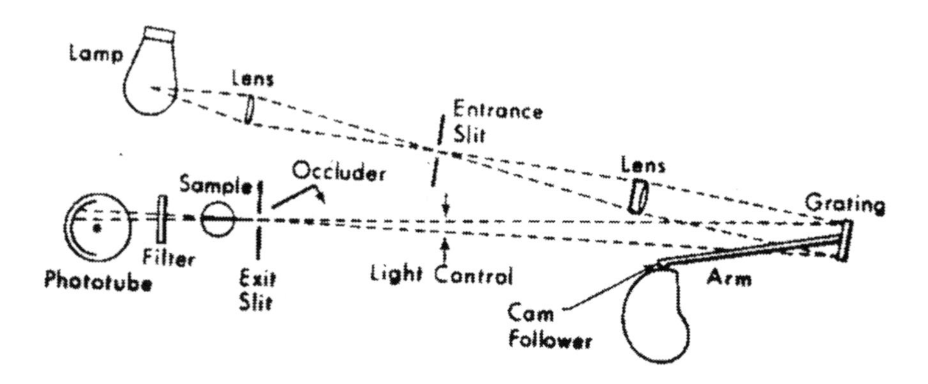

The light from a lamp is focused through a lens to an entrance slit where the light is dispersed by a rotating cam toward the exit slit whereby a narrow band of wavelengths can be absorbed or transmitted by a sample that contains a specific chemical substance. Note the location of an occluder located just before the exit slit. Also, note the role of this device in the following troubleshooting section.

3.2.5 Product Line History of the *Spec 20*

The original *Spec 20* was developed by Bausch & Lomb in 1953. The product line was sold to Milton Roy in 1985. Milton Roy sold its instrument group to Life Sciences International, renamed Spectronic Instruments Inc. in 1995. Spectronic Instruments was purchased by Thermo Optek in 1997, renamed Spectronic-Unicam in 2001, and renamed Thermo-Spectronic in 2002. In 2003 the product was moved to Madison,

WI, and the brand was renamed Thermo Electron. In 2006 Thermo Electron merged with Fisher Scientific to become Thermo Scientific. *Spec 20* instruments found in laboratories today may bear any of these brand names.

3.2.6 Troubleshooting the *Spec 20*

The troubleshooting guide shown next is included in this experiment to give the student unique insights into what can go wrong when attempting to operate this or any other analytical instrument. Analytical chemists and chemical technicians working in analytical laboratories are continuously confronted with the day-to-day reality that analytical instruments will fail, will not work properly, cannot meet QA/QC minimum criteria, etc. This table assists the student in identifying problems with the Spec 20 and how to solve these operational problems. This aspect of learning to work with analytical instruments is called troubleshooting.

Operator's Troubleshooting Guide for the Spec 20 Spectrophotometer

Problem	Possible Cause	Remedy
1. Instrument does not work	1 Power line cord not connected to outlet. 2 Dead power outlet. 3 Source lamp burned out. 4 Phototube burned out. 5 Defective electronic component.	Plug in power line cord. Try a different power outlet. Replace with new lamp. Replace as required. Refer to service manual or service center.
2. Meter does not zero	1 Sample compartment cover not closed. 2 Occluder binding. 3 Lamp access door not tightly closed. 4 Phototube defective. 5 Defective electronic component.	Close cover. Check occlude action. Close door and retighten thumbscrew. Replace as required. Refer to service manual or service center.
3. Readings drift	1 Poor sampling technique. 2 Fumes from sample. 3 Excessive line voltage variation. 4 Wrong line voltage setting. 5 Source lamp defective. 6 Phototube defective. 7 Meter defective (Spectronic 20 only). 8 Defective electronic component.	Eliminate bubbles or particles in solution. Check voltage and grounding. Reset line voltage selection switch. Replace with new lamp. Replace as required. Refer to service manual or service center. Refer to service manual or service center.

(Continued)

Operator's Troubleshooting Guide for the Spec 20 Spectrophotometer

Problem	Possible Cause	Remedy
4. Cannot set 100%T (0.0A)	1 Occluder closed.	Install test tube in sample compartment.
	2 Sample holder not fully inserted into adapter.	Insert fully.
	3 Improper filter installed.	Remove or change filter.
	4 Source lamp weak.	Replace with new lamp.
	5 Wrong line voltage setting.	Reset line voltage selection switch.
	6 Phototube weak.	Replace as required.
	7 Error in wavelength calibration.	Check calibration.
	8 Defective electronic component.	Refer to service manual or service center.
5. Readings are not repeatable even though meter reading is zero and 100%T control is set correctly	1 Loose lamp.	Tighten thumbscrew.
	2 Loose sample holder adapter.	Tighten set screw.
	3 Poor analytical technique.	Clean or replace dirty test tubes; remove bubbles, etc.
	4 Test tube position not repeating.	Always position fiducial line in exactly the sample place when test tube is inserted into adapter.
	5 Meter sticking (Spectronic 20 only).	Tap lightly for possible correction.

3.2.7 Procedure

Obtain a supply of 13 × 100 mm test tubes that are clean, dry, and free of scratches. Half-fill each tube with the 2% $CoCl_2$ solution. Set the wavelength to 510 nm on the spectrophotometer, then zero the readout. Choose one tube as a reference, place a vertical index mark near the top of the tube, and insert the tube into the sample compartment. Adjust the light control so that the meter reads 90% transmittance (T). Insert each of the other tubes and record their transmittances. If the $\%T$ is within 1% of the reference tube, place an index mark so that the tube can be inserted in the same position every time. If it is not within 1%, rotate the tube to see whether it can be brought into range. In future measurements, insert each tube in the same position relative to its index mark. Choose a set of seven tubes that have less than 1% variation in reading. Retain these tubes for subsequent photometric work and return the remainder. To compensate for variations between instruments, use the same instrument for both tube matching and experimental work.

To show that as the wavelength cam is manually rotated, different wavelengths are passed across the exit slit by the monochromator, place a piece of chalk into a test tube and insert into the sample compartment while leaving the cover open. Starting at 400 nm, scan the visible range up to 700 nm and observe the exit-slit image by looking straight down into the cell compartment as you rotate the cam.

3.3 FOR THE REPORT

Because this is an introductory experiment and learning exercise, there is no report.

Typical calibration data obtained from a visible spectrophotometer for aqueous solutions that contain dissolved cobalt chloride ($CoCl_2$) is shown in the following table. $CoCl_2$ is a crystalline solid that readily dissolves in water to yield an aqueous solution that contains Co^{2+} and Cl^- ions. Only Co^{2+} contributes to the color of the aqueous solution.

Concentration (mg/L or ppm) CoCl2 (aq)	**Absorbance (Absorbance Units, AU)**
500	0.100
600	0.235
700	0.415
800	0.610
900	0.795
1,000	1.010

Use an Excel® spreadsheet or equivalent to find the best straight line fit through the experimental x, y data points.

At the wavelength used in the measurement, a portion of a solution whose concentration is unknown is added to a cuvette, placed in the spectrophotometer, and the absorbance measured. Assuming that the unknown solute responsible for the color (in this case the aqueous cobalt[II] ion) is the same chemical species in which the previously prepared calibration plot was made, *find the concentration in #mg/L or ppm* from the calibration plot constructed, if the absorbance measured is 0.755 AU.

The graph shown below is a plot of the % absorption (not absorbance, be careful) versus wavelength. This is the absorption spectrum for Co^{2+} in the visible region of the electromagnetic spectrum. What wavelength should be used to conduct this quantitative analysis so as to obtain maximum sensitivity in measurement?

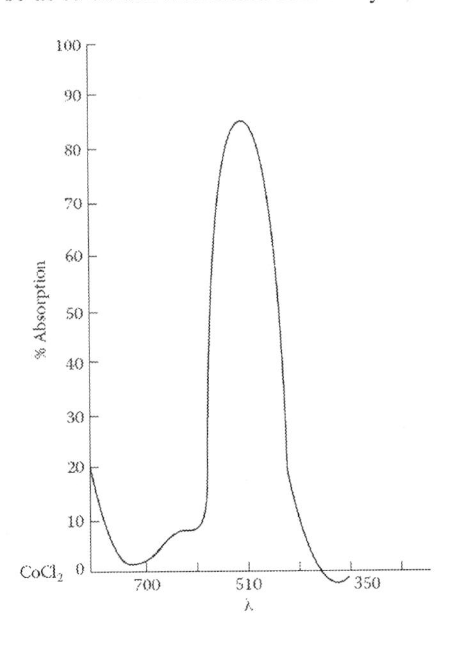

3.4 SUGGESTED READING

To develop this experiment, the author consulted the following resources:

Harris W., B. Kratochvil. *An Introduction to Chemical Analysis*. New York: Saunders College Publishing, 1981, pp. 410–412. The test matching test tube exercise was adapted from this excellent resource.

Sawyer D.,W. Heineman, J. Beebe. *Chemistry Experiments for Instrumental Methods*. New York: John Wiley & Sons, 1984, pp. 163–214.

A number of good presentations can be found on the basic principles of spectrophotometry; a particularly good presentation is found in Chapter 19 of Harris D. *Quantitative Chemical Analysis*, 3rd ed. San Francisco, CA: Freeman, 1991. Also consider Skoog D., J Leary. *Principles of Instrumental Analysis*, 4th ed. Philadelphia: Saunders, 1992. Robinson, J. *Undergraduate Instrumental Analysis*, 5th ed. New York: Marcel Dekker, 1995. A more comprehension treatment is found in: Ingle, J., S. Crouch. *Spectrochemical Analysis*. Englewood Cliffs, NJ: Prentice-Hall, 1988. Authoritative writers have joined forces and published a recent textbook that covers the broad field of chemical analysis; consider Skoog, D., J. Holler, S. Crouch. *Principles of Analytical Chemistry*, 7th ed. Boston: Cengage Learning, 2017.

4 Visible Spectrophotometric Determination of Trace Levels of Iron in Groundwater

4.1 BACKGROUND AND SUMMARY OF METHOD

Iron (Fe), a metal in great abundance in the Earth, is a common contaminant in groundwater in its oxidized forms, ferric ion, Fe^{3+} or iron (III) and ferrous ion, Fe^{2+} or iron (II). Common rust consists of ferric oxides and persists in groundwater as either solubilized or particulate matter. For environmental analytical purposes, one must distinguish between total Fe and dissolved Fe that could be present in groundwater. It is also of interest to determine the degree of metal speciation (i.e. the concentration of Fe [III] to that of Fe [II]). The hexa-aquo Fe (III) itself behaves as a weak acid and ionizes in water according to

$$Fe\left(H_2O\right)_6^{3+} \leftrightarrow Fe\left(H_2O\right)_5 OH^{2+} + H^+$$

and thus contributes to *groundwater acidity*. Pure solutions of salts that contain the hexa-aquo Fe (III) are distinctly acidic, having a pH from about 2 to 4, depending on concentration. If either total dissolved Fe or Fe (II) is to be determined, the sample must be analyzed as soon as possible after collection. If only total Fe is to be determined, the sample should be immediately acidified with hydrochloric acid.

Fe (II), once formed by chemical reduction, forms an intensely colored complex ion with 1,10-phenanthroline according to the following reactions:

DOI: 10.1201/9781003260707-4

$$Fe^{3+}_{(aq)} \xrightarrow{\text{Reducing agent}} Fe^{2+}_{(aq)}$$

Fe(II) + N N Fe(II) N N

Only the Fe (II) oxidation state for iron forms the colored complex. Hence, this selectivity provides the basis for quantitatively determining the *ferric to ferrous ion concentration ratio* that characterizes the dissolved Fe portion of the iron analysis. Several organic compounds that are readily soluble in water and easily reduce Fe (III) to Fe (II) are available and include hydroquinone, ascorbic acid, and hydroxylamine hydrochloride, among others.

This exercise affords an opportunity to introduce good laboratory practices when conducting an analysis using a UV-vis spectrophotometer. Each student is given at least one unknown groundwater sample with which to measure the concentrations of Fe (III) and Fe (II).

You will need to consider how both oxidation states of iron can be quantitatively determined using only complexation with 1,10-phenanthroline. For a given volume of groundwater, the amount of Fe (III) and the amount of Fe (II) should approximate the amount of total Fe found independently by flame atomic absorption spectrophotometry (FLAA).

4.2 EXPERIMENT

4.2.1 Volumetric Glassware Needed per Student

One volumetric flask (100 mL).
One pipette (5 mL).
One pipette (10 mL).
One volumetric flask (500 mL).
One 10 mL pipette calibrated in 1/10 mL increments (10 mL).

4.2.2 Gravity Filtration Setup

One glass funnel and standard circular filter paper.

4.2.3 Chemical Reagents Needed per Student or Group

Note: all reagents used in this analytical method contain hazardous chemicals. Wear appropriate eye protection, gloves, and protective attire. The use of concentrated acids and bases should be done in the fume hood.

15 mL of 1.0% hydroxylamine hydrochloride. Dissolve 10 g hydroxylamine hydrochloride in every 100 mL of solution.

15 mL of 0.1% 1,10-phenanthroline. Dissolve 0.1 g 1,10-phenanthroline in enough acetone until completely solubilized, then add DDI for every 100 mL of solution.

5 mL of concentrated hydrochloric acid, HCl.

0.5 g of ferrous ammonium sulfate, $Fe_2(NH_4)_2(SO_4)_3$. Dissolve 0.35 g $Fe_2(NH_4)_2(SO_4)_3$ in a 500 mL volumetric flask half-filled with DDI, then add 5 mL of concentrated HCl and adjust to the calibration mark with DDI. This gives a solution that is 100 ppm as Fe.

100 mL of saturated sodium acetate (NaOAc) solution. Dissolve enough NaOAc until crystal formation is observed.

4.2.4 Spectrophotometer

A stand-alone UV-vis spectrophotometer such as a GENESYS® (Spectronic Instruments) or equivalent is suitable.

An atomic absorption spectrophotometer (FLAA) or an inductively coupled plasma-atomic emission spectrophotometer (ICP-AES) should be available if the instructor chooses to include these instruments in this exercise. The instructional staff is responsible for operating these instruments and providing analytical results to students.

4.2.5 Procedure

1. Turn on the spectrophotometer and allow at least a 15 min warm-up time. Set the wavelength to 508 nm.
2. Prepare the calibration standards by pipetting 0 (this is called the reagent blank), 1, 2, 3, 4, and 5 mL aliquots (portions thereof) of the 100 ppm Fe stock solution into a 100 mL volumetric flask. Also, prepare an instrument calibration verification (ICV) standard by pipetting 2.5 mL of the 100 ppm Fe stock solution into a 100 mL volumetric flask. Measure the ICV's absorbance after developing the color below in triplicate. The ICV is used to evaluate the precision and accuracy of any instrumental method via interpolation of the calibration curve and is essential to maintaining good quality control. *Hint*: to minimize contamination due to carryover when using only one volumetric, *prepare standards from low to high concentration.*

Standard No.	100 ppm Fe (mL)	V (total)	Concentration (ppm)
Blank	0	—	0
1	1.0	100	1.0
2	2.0	100	2.0
3	3.0	100	3.0
4	4.0	100	4.0
5	5.0	100	5.0
ICV	2.5	100	2.5

3. To 20 mL of each reference standard solution, add 10 mL of saturated NaOAc and 10 mL of 1% hydroxylamine hydrochloride. Wait 5 min and add 10 mL of the 0.1% 1,10-phenanthroline solution. Allow 10 min, then dilute to the calibration mark of the 100 mL volumetric beaker with DDI.
4. Carefully filter approximately 35 mL of the unknown groundwater sample if necessary. Pipette 20 mL of sample and place into a clean 100 mL volumetric beaker. Add reagents as done previously for the calibration standard preparation and adjust to the final volume with DDI.
5. Set the zero control and 100% transmittance controls according to the operating manual instructions using DDI. *Record transmittance values for the blank calibration standards, ICV measured in triplicate, and unknown sample.*

4.2.6 Determination of Total Fe by FLAA or ICP-AES

You may have an opportunity to use either FlAA or ICP-AES to quantitatively measure the total iron in the same groundwater samples that were used above. The same set of calibration standards that you prepared can be used and directly aspirated into either the flame or the plasma. If you choose FlAA, you may need to install a Fe hollow-cathode lamp.

4.3 FOR THE NOTEBOOK

Use a spreadsheet program such as Excel® or its equivalent to construct a six-point calibration plot. The plot should show absorbance values on the ordinate and concentration as #ppm Fe on the abscissa. Use a least-squares regression to fit the experimental points and calculate the correlation coefficient. Calculate the percent error and the confidence interval at 95% probability for the ICV. Interpolate the calibration plot and obtain a concentration for Fe (III) and Fe (II) from the visible spectrophotometer. Obtain the concentration, C_{Fe}, for Fe(total) from the AA spectrophotometer. From these results, compare the calculated the $C_{\mathrm{Fe(II)}}/C_{\mathrm{Fe(III)}}$ ratio from the colorimetric method against the AA method (i.e. conduct a mass balance).

4.4 SUGGESTED READING

To develop this experiment, the author consulted the following resources:

Pietryzk D., C. Frank. *Analytical Chemistry*, 2nd ed. New York: Academic Press, 1979, pp. 656–657.

Annual Book of ASTM Standards Part 31, Water, 1980. Philadelphia, PA: American Society of Testing Materials, 1980, pp. 438–442.

5 Spectrophotometric Determination of Phosphorus in Eutrophicated Surface Water

5.1 BACKGROUND AND SUMMARY OF METHOD

The persistence of phosphates in lakes, rivers, and streams due to domestic and industrial pollution has led to elevated levels and is regarded as largely responsible for lake eutrophication. This was an even more serious problem up until some 20 years ago, prior to the ban on phosphate-containing detergents. In considering an environmental sample, one must distinguish between several chemical forms that contain the element phosphorus. Separation into dissolved and total recoverable phosphorus depends on filtration through a 0.45 μm membrane filter. Dissolved forms of phosphorus (P) include meta-, pyro-, and tripolyphosphates. The visible spectrophotometric procedure requires that all forms of P be chemically converted to the water-soluble orthophosphate ion, PO_4^{3-}. Hence, an acid hydrolysis step must be included in the method, which should account for all hydrolyzable P. A more rigorous conversion is necessary to include organo-phosphorous compounds and involves acid-persulfate digestion of the sample. The following flowchart summarizes the analytical scheme to differentiate the various chemical forms of phosphorus:

DOI: 10.1201/9781003260707-5

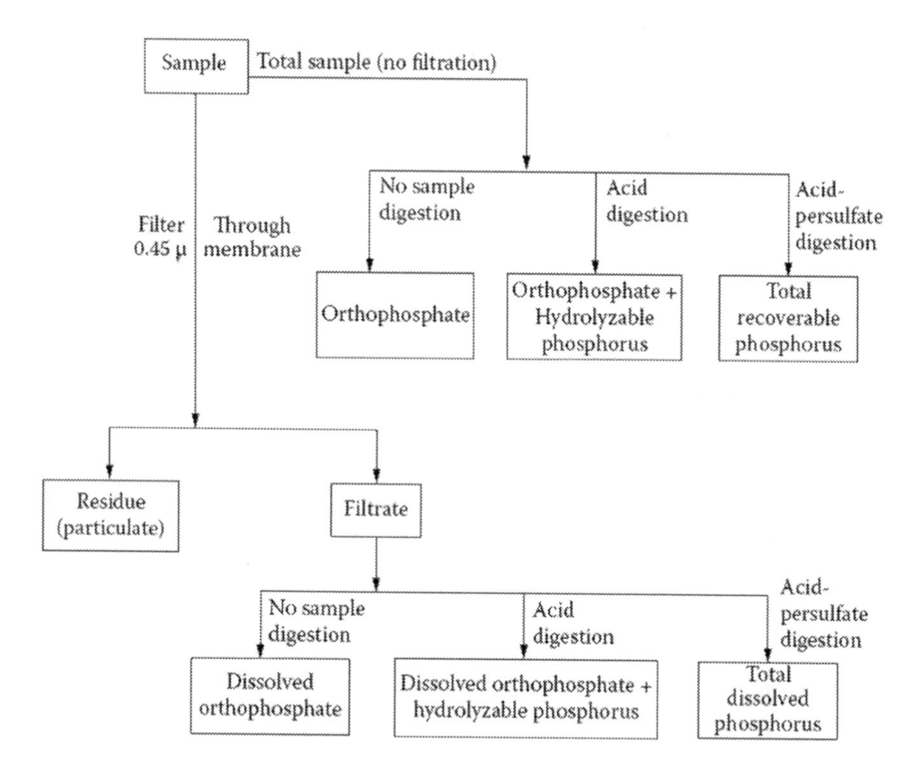

Orthophosphate forms a complex with the molybdate ion in the presence of a reducing agent such as hydrazine sulfate, amino-naphthol-sulfonic acid, tin (II) chloride, or ascorbic acid, which is commonly called *heteropoly blue* and has the molecular formula $H_3PO_4(MoO_3)_{12}$. There are two wavelengths that can be used: one between 625 and 650 nm and a more sensitive one at 830 nm. The Spectronic 21 DUV® is capable of measuring up to 1,000 nm, whereas the Spectronic 20® can go only up to 600 nm with the standard phototube. An infrared phototube is necessary to widen the wavelength of this instrument.

This exercise provides an opportunity to reinforce the principles of spectrophotometry and its relationship to environmental analysis. To minimize laboratory time, the acid hydrolysis step will be eliminated, and thus only dissolved orthophosphate, PO_4^{3-}, will be measured. In addition to determining the phosphorus content of a surface water sample, each student will be given an unknown sample by the instructor whose phosphorus concentration has been previously determined by ion chromatography.

5.2 EXPERIMENT

5.2.1 Preparation of Chemical Reagents

Note: all reagents used in this analytical method contain hazardous chemicals. Wear appropriate eye protection, gloves, and protective attire. The use of concentrated acids and bases should be done in the fume hood.

5.2.1.1 5 *M* Sulfuric Acid

Use an appropriate size graduated cylinder and add 70.25 mL of concentrated sulfuric acid to a 250 mL volumetric flask that has been previously half-filled with distilled deionized water (DDI). This solution will release heat. Allow to cool; then fill to the calibration mark with DDI. Label as appropriate. The unused 5 *M* sulfuric could be saved, diluted, and used in the exercise "Determination of Anionic Surfactants ..."

5.2.1.2 Molybdate Reagent

Dissolve 12.5 g of ammonium molybdate, $NH_4Mo_7O_{24}$, in approximately 100 mL of DDI in a 500 mL volumetric flask. Add 150 mL of 5 *M* sulfuric acid. Fill to the calibration mark with DDI. Label as appropriate.

5.2.1.3 1% Ascorbic Acid

Dissolve 1 g of ascorbic acid into a 100 mL volumetric flask half-filled with DDI. Adjust to the calibration mark with DDI. Label as appropriate.

5.2.1.4 Preparation of Stock Phosphorus

Dissolve 0.2197 g of potassium dihydrogen phosphate, KH_2PO_4, that has been dried for 1 h at 104° C in approximately 200 mL of DDI in an appropriately sized beaker. You will need the use of a correctly calibrated analytical balance. After dissolution is complete, transfer the solution to a 1 L volumetric flask and adjust to a final volume with DDI. Label as "1.0 mL = 0.05 mg P." Carefully pipette 50 mL of the stock solution into a 1 L volumetric flask half-filled with DDI. Adjust to the calibration mark with DDI. Label as "1.0 mL = 0.0025 mg P."

Sample	Milliliters Std. P (1 mL = 0.0025 mg)	Milliliters Molybdate Reagent	Milliliters Ascorbic Acid Reagent	VT (mL)	Concentration (ppm)
Blank	0	5	3	50	0
Std. 1	1	5	3	50	0.05
Std. 2	2	5	3	50	0.10
Std. 3	4	5	3	50	0.20
Std. 4	7	5	3	50	0.35
Std. 5	10	5	3	50	0.50
ICV	5	5	3	50	0.25 (expect)

5.2.2 Procedure

A 5 mL aliquot of sample is taken and transferred to a 50 mL volumetric flask. Add 5 mL of the molybdate reagent and 3 mL of the reducing agent. The mixture is diluted to volume, and after waiting 6 min for color development (time for development for standards and unknown should be the same), the absorbance is determined at 830 nm using the Spectronic 21 DUV or equivalent spectrophotometer. The ICV should be prepared in triplicate and its absorbance measured three times. Follow *good laboratory practices* (GLP), as discussed in various resources such as textbooks in analytical chemistry and in the regulatory literature.

5.3 FOR THE NOTEBOOK

Use a spreadsheet program such as Excel® or its equivalent to construct a five-point calibration plot. The plot should show absorbance values on the ordinate and concentration as #ppm P on the abscissa. Use a least-squares regression to fit the experimental points and calculate the correlation coefficient. Calculate the percent error and the confidence interval at 95% probability for the ICV. Report on the unknown #ppm P and estimate the confidence interval at 95% probability for both the surface water sample and the sample given to you by the instructor; be sure to include the code for this sample. Review this method in a resource such as *Standard Methods for the Examination of Water and Wastewater* or other sources on the colorimetric determination of phosphorus. Discuss the effect of matrix interferences on the precision and accuracy for determining P in environmental samples using the visible spectrophotometric method.

5.4 SUGGESTED READING

To develop this experiment, the author consulted the following resources:

Pietryzk D., C. Frank. *Analytical Chemistry*, 2nd ed. New York: Academic Press, 1979, pp. 658. The author again drew on this excellent resource to adapt an experiment on measuring phosphorous in eggshells to measuring phosphorous in surface water based on forming the phospho-molybdate complex ($[PO4{\cdot}2MoO3]^{3-}$).

Annual Book of ASTM Standards Part 31, Water, 1980. Philadelphia, PA: American Society of Testing Materials, 1980, pp. 438–442. The author again drew on this well-known resource to help develop the above flowchart that outlines the numerous options available when attempting to quantitate phosphorous.

6 Determination of Anionic Surfactants by Mini-Liquid–Liquid Extraction (MINI-LLE) in an Industrial Wastewater Effluent Using Ion Pairing with Methylene Blue

6.1 BACKGROUND AND SUMMARY OF METHOD

Synthetic detergent formulations make their way into the environment via industrial waste effluent. It is important that levels of these surfactants or surface-active substances be monitored. The classical analytical method that utilizes a visible spectrophotometer involves a consideration of the chemical nature of the surfactants. Surfactants are classified as either *anionic*, *nonionic*, or *cationic*, depending on the nature of the organic moiety when the substance is dissolved in water.

This mini-scale extraction method utilizes the ability of *anionic surfactants to form ion-pair complexes with cationic dyes* such as methylene blue. These complexes behave as if they are neutral organic molecules. These ion pairs are easily extracted into a nonpolar solvent, thus imparting a color to the extract. The intensity of the color becomes proportional to the concentration of the surfactants in accordance with the Beer–Lambert law of spectrophotometry.

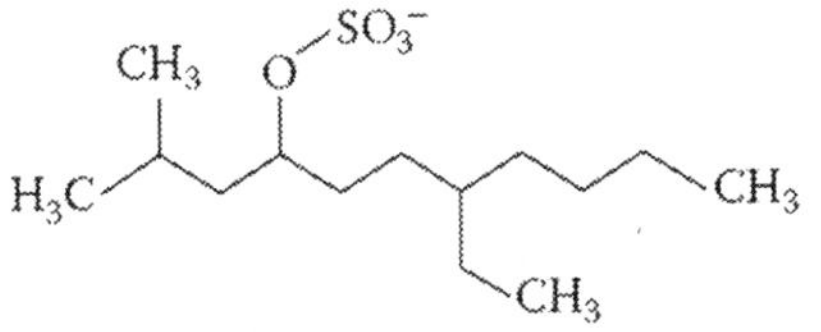

CH_3 O SO_3^- H_3C CH_3 CH_3

Anionic surfactant

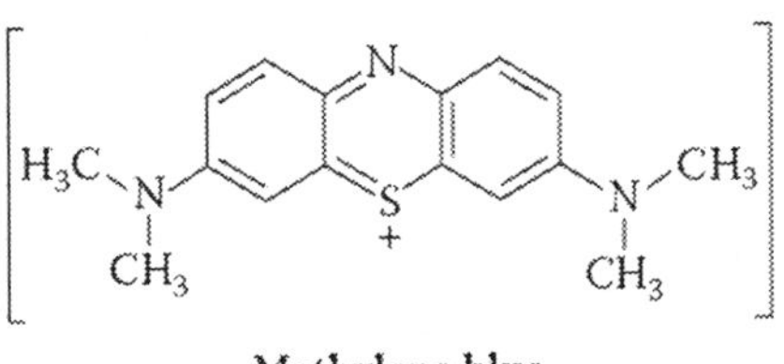

H_3C N CH_3 N S + N CH_3 CH_3 CH_3

Methylene blue

DOI: 10.1201/9781003260707-6

6.2 EXPERIMENT

This exercise introduces many of the techniques that are required to identify and quantitate an environmental pollutant while facilitating an understanding of the relationship between chemical principles and instrumental analysis.

6.2.1 Preparation of Chemical Reagents

Note: all reagents used in this analytical method contain hazardous chemicals. Wear appropriate eye protection, gloves, and protective attire. The use of concentrated acids and bases should be done in the fume hood.

6.2.1.1 Methylene Blue (MB)

Dissolve 0.05 g of MB in 50 mL of distilled deionized water (DDI).

6.2.1.2 3 *M* Sulfuric Acid

Add 16.7 mL of concentrated sulfuric acid, H_2SO_4, to a 100 mL volumetric flask half-filled with DDI. Adjust to the calibration mark with DDI.

6.2.1.3 To Prepare a 0.5 *M* Sulfuric Acid Solution

Add approximately 4 mL of the 3 *M* sulfuric acid to a 25 mL volumetric flask half-filled with DDI, then adjust to the final volume. Transfer to a storage vial and label.

6.2.1.4 To Prepare a 0.1 *M* Sodium Hydroxide Solution

Weigh approximately 0.1 g of NaOH pellets into a 50 mL beaker that contains approximately 20 mL of DDI. Dissolve with stirring. Transfer the contents of the beaker to a 25 mL volumetric flask and adjust to the final volume. Transfer to a vial and label.

6.2.1.5 To Prepare the Wash Solution

To a 1,000 mL volumetric flask half-filled with DDI, add 41 mL of 3 *M* sulfuric acid to 5 g of $Na_2HPO_4 \bullet H_2O$. Adjust to the mark with DDI.

6.2.1.6 To Prepare the MB Reagent

To a 500 mL volumetric flask half-filled with DDI, add the following:

15 mL of MB.
20.5 mL of 3 *M* sulfuric acid.
25 g of $NaH_2PO_4 \bullet H_2O$.

Shake until dissolved. Then adjust to the calibration mark with DDI. Transfer contents of the volumetric flask to a glass storage bottle and label "MB Reagent."

6.2.2 Preparation of the 100 ppm Surfactant Stock Solution and General Comments on Standards

Dissolve approximately 0.01 g of sodium lauryl sulfate in between 5 and 10 mL of MeOH in a 50 mL beaker. Transfer the contents of the beaker to a 10 mL volumetric flask and adjust to the mark with MeOH. This yields a stock solution whose concentration is 1,000 ppm. Transfer 1 mL using a glass pipette and pipette pump to a 10 mL volumetric flask. Adjust to the mark with DDI. This yields a primary dilution reference standard whose concentration is 100 ppm. Refer to the calibration table below to prepare the blank, calibration standards, and initial calibration verification (ICV) standard. A molecular structure for sodium lauryl sulfate or sodium dodecyl sulfate also known as sodium laureth sulfate is shown below:

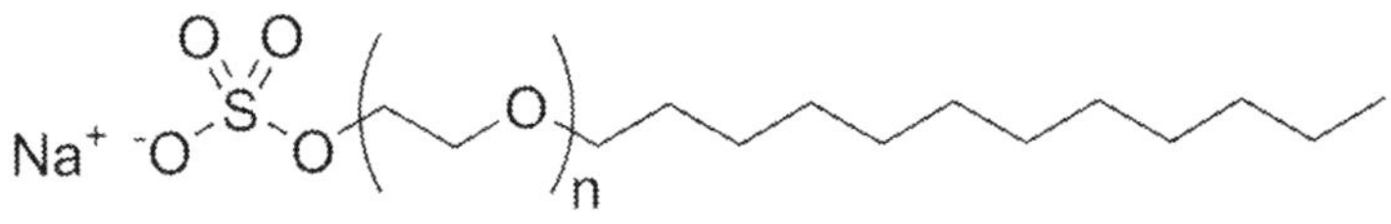

Na+ -O S O O O O n

It becomes important to *know the chemical nature of the particular surfactant* to be used to prepare the stock standard for construction of the calibration table, so that the number of parts per million (#ppm) can be related to the concentration of anionic sulfonate actually taken and extracted as an ion pair. For example, for *p*-toluene sulfonic acid, 100 mg as the *p*-toluene sulfonate ion is only 100.6 mg as the acid; however, 100 mg as the *p*-toluene sulfonate ion is 113.4 mg as its sodium salt. A stock solution is stable for no more than one week. Working solutions such as the 100 ppm surfactant should be prepared fresh daily. Keep in mind that approximately *0.1 g* of any pure solid dissolved in enough solvent to prepare *10 mL* of solution yields a standard whose concentration is approximately *10,000 ppm!*

Calibration Table

Sample	100 ppm Surfactant (#μL)	Surfactant Added (#μg)	MB Reagent (#μL)	DDI (#μL)	CH2Cl2 (total) (#μL)	Concentration Original Sample (#ppm)
Blank	0	0	2.5	10	10	0
Std.1	100	10	2.5	10	10	1.0
Std.2	250	25	2.5	10	10	2.5
Std. 3	500	50	2.5	10	10	5.0
Std. 4	1000	100	2.5	10	10	10
ICV	400	40	2.5	10	10	4.0

6.2.3 Operation and Calibration of the Orion SA 720A pH Meter

The pH meter must be set up and calibrated with two buffers. Use buffer solutions whose pH values are 7 and 10, as these buffers have pH values near those required in this method. Refer to the instructions for operating the specific model of the pH meter available at your workbench.

6.2.4 Procedure to conduct a miniaturized liquid-liquid extraction (LLE)

1. Using a clean 25 mL graduated cylinder, place 10 mL of an aqueous sample whose anionic surfactant concentration is to be determined into a 50 mL beaker. The aqueous sample could be a blank (i.e. a sample with all reagents added except for the analyte of interest), calibration standard, ICV, fortified (i.e. spiked) sample, or an actual unknown wastewater effluent sample.
2. Neutralize the sample to a pH between 7 and 8 by dropwise addition of either 1 *M* NaOH or 0.5 *M* H_2SO_4.
3. Add 2.5 mL of the MB reagent and swirl; then transfer the contents of the beaker to a 30 mL glass separatory funnel. *Be sure the stopcock is closed.*
4. Add 2 mL of methylene chloride (dichloromethane) or equivalent solvent. Because methylene chloride is much denser than water, it will comprise the lower layer after the two phases separate.
5. Stopper the separatory funnel with the ground-glass top, invert the funnel, and shake. Be sure to vent the vapor. This should be done in the fume hood.
6. Withdraw the lower layer into a second clean beaker.
7. Extract the remaining aqueous phase two more times with 2 mL portions of methylene chloride. Then combine the methylene chloride extracts. Approximately 6 mL of organic solvent should be obtained at this point.
8. Wash the combined extracts with wash solution. To do this, transfer the 6 mL of extract ($MeCl_2$ + dissolved ion pair) to a clean 30 mL separatory funnel. Add approximately 10 mL of wash solution, shake, allow time for the two phases to separate, and then remove the lower layer directly into a clean 50 mL beaker.
9. Transfer the washed extract (approximately 6 mL) to a 10 mL volumetric flask and adjust to a final volume.
10. Transfer a portion of the 10 mL methylene chloride extract to a standard spectrophotometric cuvette and measure the absorbance against methylene chloride as a blank in the reference cell at 652 nm. *Record and repeat steps 1 through 10 for all standards and samples.*
11. *Discard the waste extract into the hazardous waste receptacle* located in the laboratory.

6.3 FOR THE REPORT (A WRITTEN LABORATORY REPORT DUE ON THIS EXPERIMENT)

Use a spreadsheet program such as Excel® or equivalent to construct a four-point calibration plot. The plot should show absorbance values on the ordinate and concentration of surfactant in ppm on the abscissa. Calibration data and the least-squares regression plotted in Excel for both an external standard mode and a standard addition mode for the quantitative analysis of anionic surfactants using the methylene blue colorimetric method *are shown in the following section*. Use a least-squares regression to fit the experimental points and calculate the correlation coefficient. Calculate and report on the percent error and the confidence interval at 95% probability for the ICV. *Report* on the concentrations of any unknown environmental samples to which this method was applied. *Discuss* what you learned from this environmental analysis method drawing on your previous experience with spectrophotometric methods. *Comment* on the precision and accuracy afforded by the analytical method.

Two representative calibration plots comparing *the external standard* approach and *the standard addition* approach to instrument calibration for this analytical method are shown here:

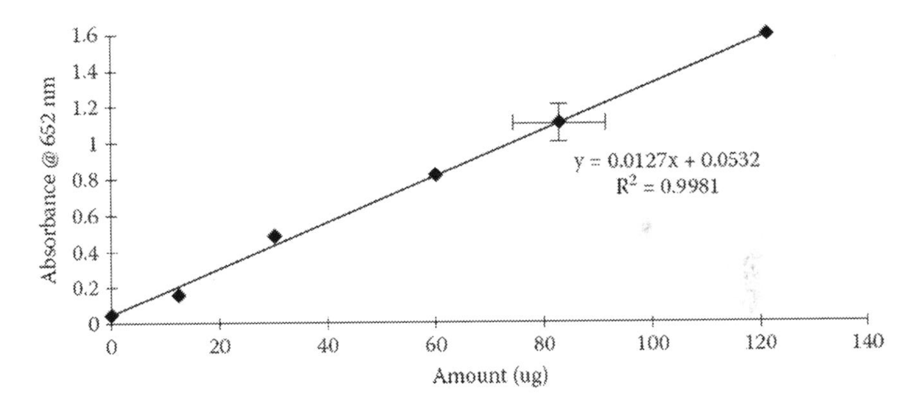

Calibration curve of anionic surfactants paired with methylene blue dye using the *external standard* mode of instrument calibration. The error bars represent the standard deviation in ICVs. ◆, standards; ■, ICV; linear regression on standards.

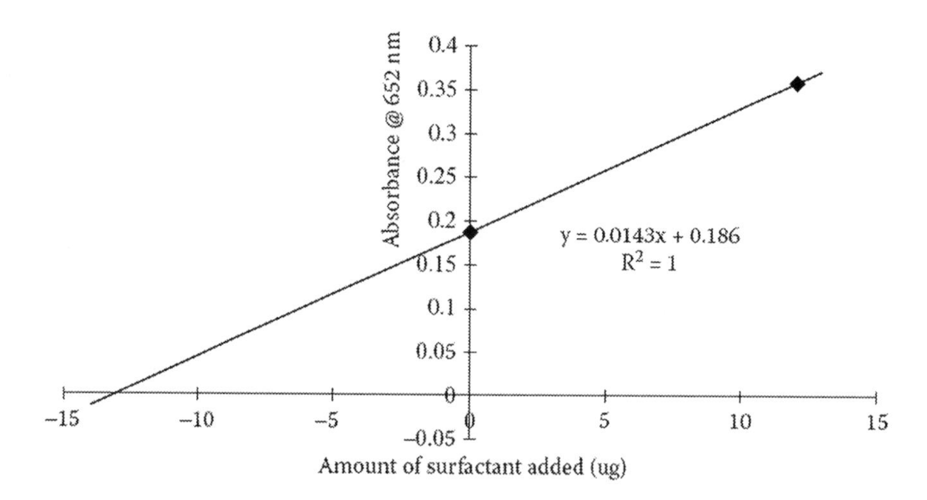

Calibration curve of anionic surfactants paired with methylene blue dye using the *standard addition* mode of instrument calibration. This graph is typical of how points are plotted when the standard addition approach to instrument calibration is implemented. The equation is for the regressed line through the points for the spiked sample and for the average of the triplicate runs of the unspiked samples.

6.4 SUGGESTED READING

To develop this experiment, the author consulted the following resources:

Sawyer D., W. Heinemann, J. Beebe. *Chemistry Experiments for Instrumental Methods.* New York: Wiley, 1984, pp. 163–214.

Schmitt T.M. *Analysis of Surfactants, Surfactant Science Series Volume 40.* New York: Marcel Dekker, 1992, pp. 290–292.

7 Comparison of Ultraviolet and Infrared Absorption Spectra of Chemically Similar Organic Compounds

7.1 BACKGROUND AND SUMMARY OF METHOD

Because most organic compounds known to pollute the environment are colorless, it would be close to impossible to identify them and therefore to quantitate if only a visible spectrophotometer were available. Do not despair! The ultraviolet (UV) spectrophotometer enables the full UV region to be used, and these organic contaminants can be identified. The UV region is defined to be those wavelengths of electromagnetic radiation *from 190 to 400 nm*. We know that the internal energies of atoms and molecules are quantized; that is, only certain discrete energy levels are possible, and the atoms and molecules must exist at all times in one or the other of these allowed energy states. For absorption of radiation to occur, a fundamental requirement is that the energy of the photon absorbed must match exactly the energy difference between initial and final energy states within the atom or molecule. Consideration of atoms falls within the realm of atomic absorption and atomic emission energy states. In contrast, *molecular absorption* involves transitions between electronic (ultraviolet-visible absorption), vibrational (mid-infrared absorption), and rotational (microwave absorption) quantized energy states. The following diagram depicts quantized energy states for organic molecules that are dissolved in a solvent:

DOI: 10.1201/9781003260707-7

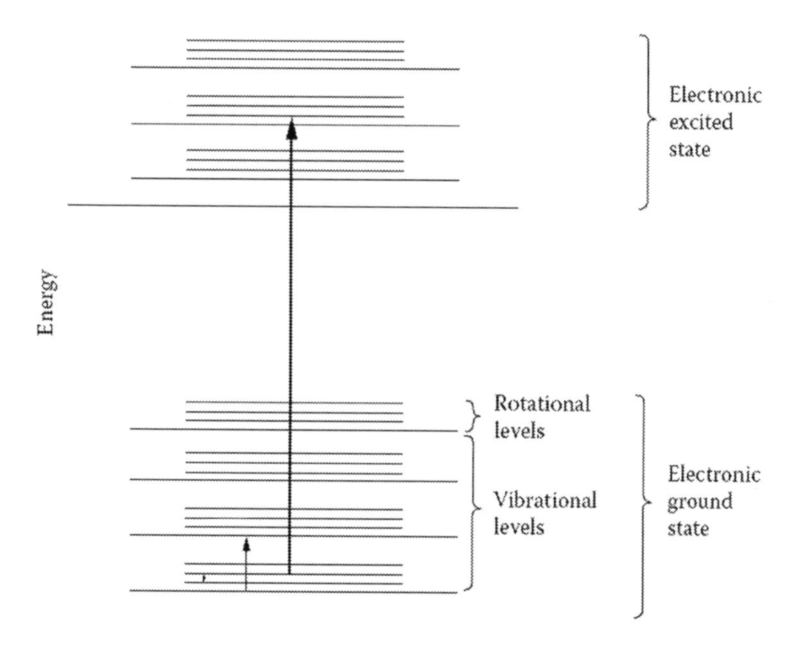

The molecular absorption phenomenon can only be accurately measured provided that the ratio of transmitted intensity, I, of the UV radiation to that of the incident intensity, I_0, is due to the presence of the dissolved solute and not to scattering of the incident beam. If the optical windows (see the following section) absorb UV radiation, then the absorbance that is related to the logarithm of the ratio I/I_0 would cause an increase in sample absorbance; hence, this would lead to an erroneous result. The student will encounter two types of optical window material. One consists of *glass* and is said to have a UV cutoff (UV wavelengths below the cutoff value would absorb) of 300 nm (near UV), and the other consists of *quartz* with a UV cutoff of 190 nm. The rectangular cuvette is depicted as follows:

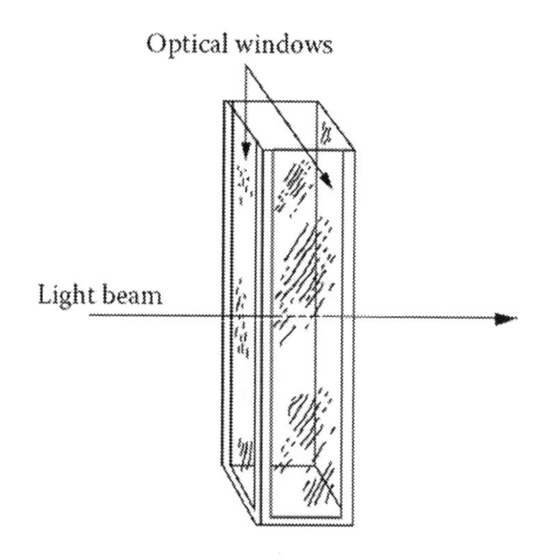

7.1.1 UV-Vis Absorption Spectroscopy

The light source for molecular absorption spectroscopy in the visible and UV regions (i.e. 750 nm down to 350 nm) is a tungsten filament lamp, which radiates as a black body at about 2,800° K. Below 350 nm, the hydrogen or deuterium gas discharge lamp is preferred. The *reciprocal linear dispersion* (refer to "Suggested Reading" below or other equivalent text for explanation) is around 1 nm/mm for diffraction grating spectrophotometers and between 0.5 and 5 nm/mm for prism spectrophotometers. (*Note*: dispersion depends on wavelength for a prism.) Air absorbs UV light of wavelengths shorter than about 180 nm, so studies at wavelengths shorter than 180 nm require the use of an evacuated spectrometer. This region is thus termed the *vacuum ultraviolet*.

The role of the solvent becomes critical in obtaining accurate UV absorption spectra. A solvent considered suitable for use in the UV-vis region must itself exhibit a low absorbance as well as dissolve the solute whose spectrum is sought experimentally. Fortunately, most common solvents are highly transparent to visible light, but all begin to absorb at some wavelength in the UV. It is not essential that the solvent have 100% transmittance, although it is desirable that as large a fraction as possible of the incident radiant energy be available for absorption by the solute. The following table lists the *approximate absorption cutoffs* for several widely used solvents. The cutoff defines a practical short-wavelength (λ) limit for the useful range of a solvent. Below this wavelength, the absorbance of the solvent, if placed in a 1 cm cell, exceeds 1.0 absorbance unit full scale (AUFS).

Solvent	Cutoff λ (nm)	Solvent	Cutoff λ (nm)
Water	200	Dichloromethane	230
Hexane	200	Chloroform	245
Heptane	200	Carbon tetrachloride	265
Cyclohexane	210	Dimethyl sulfoxide	265
Methanol	210	Dimethyl formamide	270
Ethanol	210	Benzene	275
Acetonitrile	190	Pyridine	300
Dioxane	215	Acetone	325

Different solutes exhibit different UV absorptivities. Recall that what is measured in a spectrophotometer is absorbance, and absorbance A is related to absorptivity ε via the following mathematical relationship:

$$A = \varepsilon bc$$

Two chemically different solutions that contain solutes at identical concentrations in a suitable solvent using the same cuvette would not be expected to have the same absorbance due to *differences in absorptivity* (earlier terms included molar absorptivity and extinction coefficient). If logarithms are taken on both sides of the above equation, we have

$$\log A = \log \varepsilon + \log bc$$

Because the log bc term is independent of wavelength, log A will vary only as a function of absorptivity. Thus, a plot of log A vs. spectrophotometer wavelength setting will be the same even though concentrations and path lengths for individual samples may differ. In this way, a comparison of the different curves can be made.

7.1.2 Mid-Infrared Absorption Spectroscopy

The mid-infrared region of the electromagnetic spectrum begins on the higher energy end at 2.5 μm (4,000 cm^{-1}) and ends on the lower energy end at 6 μm (650 cm^{-1}). Dispersive type infrared absorption spectrometers have given way to *Fourier-transform infrared (FTIR) spectrometers*. Refer to "Suggested Reading" or other equivalent texts to better understand the principles underlying infrared spectroscopy.

7.2 EXPERIMENT

This exercise affords students an opportunity to measure actual UV and FTIR absorption spectra of several organic solutes while comparing overall differences in UV and FTIR spectra for chemically similar and dissimilar organic solutes. The laboratory experiment focuses on the influence of delocalization of electron density and the nature of UV absorption spectra. The experiment also focuses on the relationship between organic functional group analysis and FTIR absorption spectra.

Two sets of solute/solvent chemical systems are to be studied in this exercise. The first set consists of *organo-sulfonates* and enables a comparison of UV absorption spectra of alkane vs. aromatic sulfonates that are dissolved in water. The second set consists of two representative *organic esters* that differ in their carbon backbone. The student is to engage both the UV spectrophotometer and the FTIR spectrometer in accomplishing the experimental aspects of this exercise.

7.2.1 Items/Accessories Needed per Student or Group

One pair of matching rectangular quartz cuvettes that transmit between 200 and 350 nm.

One polyethylene (PE) disposable IR card (Type 61, 3M).

One polytetrafluoroethylene (PTFE) disposable IR card (Type 62, 3M).

One model 160-UV (Shimadzu) UV-vis scanning absorption spectrophotometer or another equivalent scanning instrument.

One model 1600 Fourier-transform infrared spectrometer (PerkinElmer) or another equivalent instrument.

7.2.2 Preparation of Chemical Reagents

Note: all reagents used in this analytical method contain hazardous chemicals. Wear appropriate eye protection, gloves, and protective attire. The use of concentrated acids and bases should be done in the fume hood.

7.2.3 Procedure to Obtain UV Absorption Spectra for Two Sets of Chemically Similar Organic Compounds: (1) Alkane Sulfonates vs. Alkyl Sulfates and (2) Two Esters with Different Carbon Backbones

Prepare two aqueous solutions that contain approximately 1,000 ppm each of *1-octanesulfonic acid* or *sodium dodecyl sulfate* (use whichever is available) *and p-toluene sulfonic acid* dissolved in distilled deionized water (DDI). Transfer a portion of this solution into a quartz rectangular cuvette and record the UV absorption spectrum for both organic compounds. Prepare a 100 ppm solution that contains *methyl methacrylate* and a 100 ppm solution that contains *ethyl glycolate*. You will need to use the wavelength cutoff guide to choose a suitable solvent because these organic compounds do not dissolve to any great extent in water. Record the UV absorption spectrum between 200 and 350 nm. Compare the spectra when a solvent is placed in the reference beam of the dual-beam instrument. Molecular structures for these unique organic compounds are shown here:

1-octanesulfonic acid 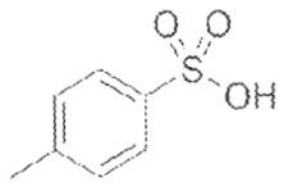 | p-toluene sulfonic acid 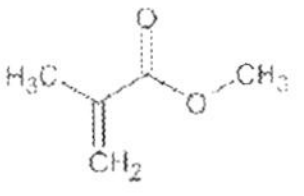 | methyl methacrylate 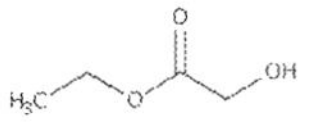 | ethyl glycolate

7.2.4 Procedure to Obtain FTIR Absorption (Transmission) Spectra for Various Organic Compounds

Disks that are partially to fully transparent in the infrared are available to easily prepare samples. An aqueous solution containing the anionic surfactants can be deposited directly onto either the PTFE or PE disk. Allow sufficient time for the solvent to evaporate off of the disk. Conduct a survey scan and take 16 (sixteen) FTIR scans. Plot the spectra. Consult the staff for assistance with the Model 1600 FTIR® (PerkinElmer). FTIR absorption (transmission) spectra for both sulfonate surfactants and for both esters can be obtained using this technique.

7.3 FOR THE REPORT

This exercise has introduced selected principles of molecular UV and infrared spectroscopy to conduct *qualitative* chemical analysis. Interpret the significance of these spectra in terms of molecular structure. You should have obtained hard-copy printouts of UV and FTIR spectra. Please include all relevant spectra in your report.

Suppose that a client sent to you a mixture that was prepared from the ethylbenzene and styrene solutions that you used in the lab. Would benzene be a suitable solvent to use in obtaining UV spectra for ethylbenzene and styrene? Discuss how you would design a *quantitative* analysis to determine the concentration of ethylbenzene and of styrene in the client's sample. Assume that you have available pure ethylbenzene and pure styrene (chemists refer to these as neat forms of the liquids).

7.4 SUGGESTED READING

To develop this experiment, the author consulted the following resources:

Crooks J. *The Spectrum in Chemistry.* New York: Academic Press, 1978, pp. 100–101.

Leary S.D.J. *Principles of Instrumental Analysis*, 4th edn. Philadelphia, PA: WB Saunders, 1992.

Sawyer D., W. Heinemann, J. Beebe. *Chemistry Experiments for Instrumental Methods.* New York: Wiley, 1984, pp. 215–221.

Thompson C. *Ultraviolet-Visible Absorption Spectroscopy.* Boston, MA: Willard Grant Press, 1974, Chapters 2 and 3.

8 Determination of Oil and Grease and of Total Petroleum Hydrocarbons in Wastewater via Reversed-Phase Solid-Phase Extraction Techniques (RP-SPE) and Quantitative Fourier-Transform Infrared (FTIR) Spectroscopy

8.1 BACKGROUND AND SUMMARY OF METHOD

One of the most informative and straightforward analytical methods involves the quantitative analysis of environmental samples that have been contaminated with what can be collectively termed oil and grease. The term *oil and grease* describes the extent to which an environmental sample such as wastewater or soil is contaminated. *Oil and grease* refers to any and all hydrocarbons including lipids, high-molecular-weight fatty acids, triglycerides, higher olefinic hydrocarbons, alkanes, alkenes, monocyclic and polycyclic aromatics, and so forth. The *Total Petroleum Hydrocarbon* (TPH) content of this oil and grease contamination can also be quantitatively determined by this method if the extracted *sample is placed in contact with silica gel*. The software that will be used with the FTIR spectrophotometer distinguishes between these two definitions. The absorption in the infrared region of the electromagnetic spectrum due to the presence of *aliphatic and aromatic carbon to hydrogen-stretching vibrations* occurs in the 3,100 to 2,900 cm^{-1} range. These concepts form the physicochemical basis for instrumental measurement.

DOI: 10.1201/9781003260707-8

The method is adapted from Method 5520, "Oil and Grease," in *Standard Methods for the Examination of Water and Wastewater* and follows from EPA Method 413.2. EPA Method 1664 Revision B and EPA Method 413.1 replaced the extraction solvent Fluorocarbon 113 or 1,1, 2-trichloro-1,2,2,-trifluoroethane with n-hexane. At present, EPA Method 1664B is called n-hexane extractible and the sample matrix is referred to as *HEM oil and grease* while a replacement for the infrared absorption determinative technique is provided by a gravimetric determinative technique following distillation and removal of the n-hexane extraction solvent.

Primarily for educational purposes, the FTIR determinative technique will be used in this experiment because it enables students to again be introduced to the *reversed-phase solid-phase extraction sample prep technique*. We will *responsibly recover* the Fluorocarbon 113 solvent whose waste disposal is the subject of current environmental concern and debate. Software developed at PerkinElmer Corporation during the late 1980s has been downloaded to the Model 1600® FTIR spectrophotometer and will be used to provide for the quantitative analysis. The original concept was first published by and at the EPA in the early 1970s.

An aqueous sample whose oil and grease content is to be determined is first acidified, then extracted using 1,1,2-trichloro-1,2,2-trifluoroethane (TCTFE) which is a liquid at room temperature. TCTFE has a *normal boiling point* of 48° C. This relatively low boiling point classifies this solvent as a VOC. The molecular structure for this solvent is shown below:

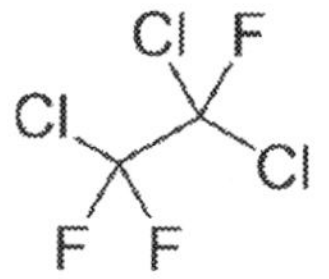

Cl F
Cl
Cl
F F

1,2,2-trichloro-1,2,2-trifluoroethane

Molecules of this solvent lack a C–H covalent bond and thus serve to provide an excellent background infrared spectrum because no absorption in the 3,100 to 2,900 cm^{-1} region is found. The extract is isolated from the aqueous matrix, and an aliquot (a portion thereof) is transferred to a 1 cm (path length) quartz cuvette. The FTIR absorbance vs. wavelength or wave number is graphically displayed on the screen. This absorbance, which is initially related to the concentration of oil and grease in the chemical reference standards used to calibrate the instrument yields the concentration as oil and grease in the unknown sample via interpolation of the least-squares regression fit. The common concentration unit used is mg of oil and grease per 100 mL of solution (reported as a #mg oil and grease/100 mL sample). Students should be aware that the reference standards and samples after RP-SPE will be dissolved in TCTFE.

A soil/sediment or contaminated sludge with a very high solids content can be extracted via Soxhlet extraction techniques or, more conveniently, via ultrasonic probe sonication into a mixture of TCTFE and a carefully weighed amount of soil/sediment/sludge.

8.2 EXPERIMENT

8.2.1 Preparation of Chemical Reagents

Note: all reagents used in this analytical method contain hazardous chemicals. Wear appropriate eye protection, gloves, and protective attire. The use of concentrated acids and bases should be done in the fume hood.

8.2.2 Reagents Needed per Student or Group of Students

0.1 g of hexadecane, $C_{16}H_{34}$. Alternatively, dodecane, $C_{12}H_{26}$, can be substituted.
0.1 g of iso-octane (2, 2, 4-trimethyl pentane), $(CH_3)_3CCH_2CH(CH_3)$.
0.1 g of benzene, C_6H_6.
250 mL of TCTFE.
100 mL of MeOH (methanol). Use as high a purity as is available.
5 g of silica gel, needed only if TPHs are to be determined in addition to oil and grease.
5 g of sodium sulfate anhydrous (Na_2SO_4).
20 mL of 1:1 hydrochloric acid (HCl). Mix equal volumes of acid and deionized water. Remember, *always add acid to water*.

8.2.3 Apparatus Needed per Group

Vacuum manifold to conduct solid-phase extraction (SPE).
Water trap and associated vacuum tubing to be used with the vacuum manifold.
Suction vacuum pump connected to the water trap.
Accessories for use of the SPE vacuum manifold, including a receiving rack.
SPE cartridges packed with C_{18}-bonded silica.
70 mL sample reservoirs.
10 mL glass volumetric flasks.

8.2.4 Procedure

Refer to the oil and grease analysis method from *Standard Methods* (refer to "Suggested Reading") or equivalent and implement the appropriate procedure. For wastewater samples, follow the procedure outlined below. If a percent recovery study is required, the procedure that immediately follows provides guidance for this study.

8.2.5 Percent Recovery Study

To isolate and recover TPHs from water by RP-SPE combined with quantitative FTIR:

- Spike approximately 200 mL of high-purity laboratory water that has been *acidified with 1:1 HCl* with approximately 2 mg of dodecane; note that most of the dodecane will float on top. This sample is called a *matrix spike*, using

terms first developed at the EPA. Spike a second 200 mL portion of *acidified* water with approximately 2 mg of dodecane. This sample is called a *matrix spike duplicate*. Leave a third 200 mL portion of water unspiked and *acidify*. This sample is called a *method blank*.
- Set up three C_{18} SPE cartridges and condition with MeOH.
- Pass the method blank, matrix spike, and matrix spike duplicate samples through the cartridges under vacuum.
- Remove water droplets with a Kimwipe; apply more vacuum to remove water from the sorbent.
- Elute off the cartridge with two 500 μL portions of TCTFE into a 1 mL volumetric as receiver; use a glass syringe that is available near the SPE manifold.
- Transfer contents of the 1 mL receiver to the quartz cuvette and add exactly 1.0 mL of TCTFE using a glass pipette.
- Call previous calibration standards from disk on Model 1600 FTIR.
- Run each of the eluents from SPE. Transfer each eluent to a quartz cuvette and measure the absorbance. For the sample ID, enter any number. For the initial mass of the sample, enter "200," and for the volume of the sample after extraction, enter "2."

For the preparation of a control (i.e. a 100% recovered sample), weigh approximately 2 mg of dodecane into a 1 mL volumetric flask half-filled with TCTFE and adjust to the mark with TCTFE. Transfer to a quartz cuvette, add 1 mL of TCTFE, and measure the absorbance using the Model 1600® PerkinElmer FTIR or equivalent instrument.

8.2.6 Probe Sonication: Liquid–Solid Extraction

It is first necessary to estimate to what extent the solid sample is laden with oil and grease. This minimizes the necessity to dilute the extract so that the absorbance will remain on scale. This can be accomplished by taking 0.5, 5, and 15 g samples and using identical extraction volumes. Once the optimum sample weight has been estimated, proceed to the next step.

8.2.7 Calibration of the FTIR Spectrophotometer

Calibrate the Model 1600® FTIR (PerkinElmer) or equivalent instrument by first preparing a series of working calibration standards. The table below serves as a useful guide and yields concentrations that are compatible with the software that operates the instrument.

Prepare Blend A by obtaining a total weight of approximately 0.10 g for the pure form of the oil that is to be defined as the reference. For example, if hexadecane, isooctane, and benzene are to be used and mixed, add approximately 0.033 g of each to obtain the desired weight. Transfer the oil to a clean, dry 10 mL volumetric flask. Add about 5 mL of TCTFE to dissolve the oil, then adjust to the calibration mark. Transfer to a clean, dry glass vial with a Teflon/silicone septum and screw cap. Label this solution "Blend A, 1,000 mg Oil and Grease/100 mL (in TCTFE)"; prepare Blends B through F according to the following table:

Blend	Blend A (mL)	Extract Volume (mL)[a]	Concentration (mg/100 mL)
F	0.01	10	1.0
E	0.1	10	10
D	0.2	10	20
C	0.3	10	30
B	0.4	10	40
ICV[b]	0.25	10	Unknown

[a] Use 10 mL glass volumetric flasks and TCTFE to adjust to final volume.

[b] ICV = instrument calibration verification standard; run as if it were a sample; enter a "1" for sample weight and "1" for extract volume.

When you are ready to perform the calibration, retrieve under method "og & ph" (oil, grease, and petroleum hydrocarbons) and exercise one of the six options. It is important to obtain a fresh background by placing TCTFE into a clean 1 cm quartz cuvette. If the blank reveals a large absorbance, the *quartz cuvettes must be cleaned with detergent*. Contamination of the surface of the quartz cuvettes represents a major source of error with this method.

8.2.8 Isolation, Recovery, and Quantitation of Oil and Grease from Wastewater Samples

- Place approximately 200 mL of wastewater sample into a clean, dry 250 mL beaker using a graduated cylinder. *Record the volume of sample in your lab notebook*.
- Acidify to a pH of approximately 2 by adding sufficient 1:1 HCl.
- Set up the SPE vacuum manifold and condition the sorbent with MeOH; the sorbent surface should be wet with MeOH prior to passing the wastewater sample through the cartridge.
- Connect the 70 mL polyethylene sample reservoir to the cartridge via the adapter.
- Pass 200 mL of sample through the manifold under vacuum; observe that the sample actually flows through the SPE sorbent and watch for plugging.
- Remove the reservoir; remove the water droplets on the inner wall of the cartridge barrel.
- Elute with two 500 μL of TCTFE directly into a 1.0 mL volumetric flask.
- Add enough anhydrous sodium sulfate, if necessary, to remove residual water in the eluent and stir.
- Transfer the contents of the volumetric flask to the quartz cuvette, add, with a pipette, 1 mL of additional TCFFE, and *measure the absorbance* using the Model 1600 FTIR.

- The printer will give you a hard-copy output after a certain number of FTIR scans have been acquired. *Record all absorbance data in your laboratory notebook.* Transfer all waste TCTFE to a properly labeled waste receptacle for recovery purposes.
- If the absorbance is too large, make the appropriate dilution. Repeat the dilution step if necessary. Transfer all waste TCTFE to a properly labeled waste receptacle for recovery purposes.
- Prior to each day's FTIR measurements, the ICV standard should be measured and a log kept of its daily interpolated concentration. If the ICV value becomes a statistical outlier, rerun the calibration.

8.3 CALCULATIONS

The printout for a correctly measured TCTFE extract gives a direct value for the concentration of oil and grease in a solid sample, provided the weight of the sample is entered in grams and the volume of the extract is entered in milliliters. Because we are measuring sample volume, instead of weight, *substitute for the weight with 200 mL* and use an extract volume of *2 mL*. The calculation is based on the following:

$$\frac{(2mL)(\#\,mg\,oil\,and\,grease\,/\,100mL)(DF)}{200mL\,sample} \times \frac{1000mL}{1\text{L}}$$

$$= \#\frac{mg}{L}(oil\,and\,grease)\,in\,wastewater\,sample$$

where DF is the dilution factor. If no dilution, assume DF = 1. If a 1:25 dilution is made, DF would equal 25.

8.4 SUGGESTED READING

To develop this experiment, the author consulted the following resources:

Eaton A., L. Clesceri, A. Greenburg, Editors. *Standard Methods for the Examination of Water and Wastewater*, 19th edn. Washington, DC: American Public Health Association, American Water Works Association, Water Environment Federation, 1995, pp. 5-30–5-35.

EPA Method 413.2, Oil and Grease. *Total Recoverable Hydrocarbons*, Test and Evaluation Facility, Environmental Protection Agency, Cincinnati, OH, 1978.

Method 1664, Revision B: n-Hexane Extractable Material (HEM; Oil and Grease) and Silica Gel Treated n-Hexane Extractable Material (SGT-HEM; Non-polar Material) by Extraction and Gravimetry Revision B. February 2010.

McGratton, B. *Method Program Automating ASTM Standard Test Method for Oil and Grease and Petroleum Hydrocarbons in Water.* PerkinElmer Infrared Bulletin 114, 1600.

Method 1664, N-Hexane extractable Material by Extraction and Gravimetry, EPQ-821-B-94-004. Washington, DC: Office of Water Engineering and Analysis Division, 1995.

This experiment was adapted from Gruenfeld, M. *Enviro Sci Technol* 7: 636–639, 1973.

9 Determination of the Degree of Hardness in Various Sources of Groundwater Using Flame Atomic Absorption Spectroscopy

9.1 BACKGROUND AND SUMMARY OF METHOD

The extent to which groundwater has been rendered *hard* has been defined as the *concentration of dissolved bicarbonates containing calcium (Ca^{2+}) and magnesium (Mg^{2+}) ions present in the sample*. Hardness can be quantitatively measured by finding some way to measure these two alkaline–earth metal ions. The classical method that still finds widespread use is titration with EDTA. We are going to approach the problem by measuring the concentration of Ca^{2+} by *flame atomic absorption spectroscopy* (FLAA) and in addition, by a mere change of hollow-cathode lamps, by measuring the concentration of Mg^{2+} by FLAA. Several sources of groundwater will be obtained, and the concentration of the chemical elements Ca and Mg will be used to estimate the degree of water hardness.

Recall that an *analytical method's precision* is a measure of the degree to which it can be reproduced or repeated and is evaluated by calculating the standard deviation for the method's instrument calibration verification (ICV) standard following the establishment of a single-point or multipoint calibration. An *analytical method's accuracy* is a measure of how close the results are to an established or authoritative value and is evaluated by calculating the percent relative error.

FLAA requires a means by which an aqueous solution containing metal ions can be aspirated into a reducing flame environment by which atomic Mg or Ca vapor is formed. Photons from the characteristic Mg emission of a hollow-cathode lamp (HCL) are absorbed by ground-state Mg atoms present in the approximately 2,300° C air–acetylene flame. The amount of radiant energy absorbed as a function of the concentration of an element in the flame is the *basis of quantitative analysis using AA* and follows Beer's law. In contrast to molecular absorption in solution, atomic spectra consist of lines and originate from either atomic absorption or atomic emission processes, which are depicted schematically here:

DOI: 10.1201/9781003260707-9

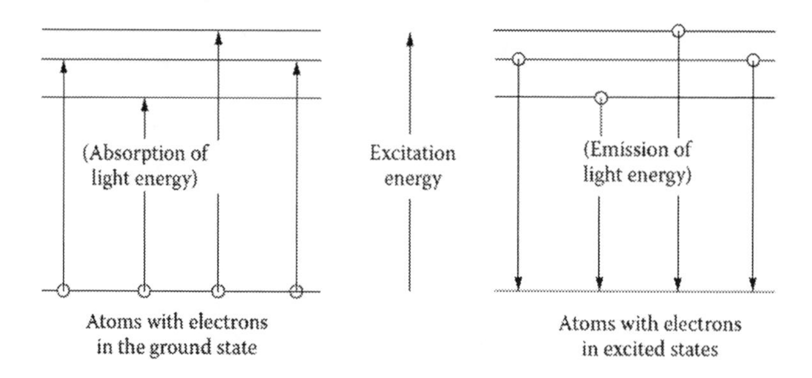

Instrument detection limits (IDLs) for most metals by FLAA are in the low-ppm realm in contrast to graphite furnace AA (GFAA) where IDLs can reach down to low-ppb concentration levels. The conventional premixed chamber-type nebulizer burner is common. The sample is drawn up through the capillary by the decreased pressure created by the expanding oxidant gas at the end of the capillary, and a spray of fine droplets is formed. The droplets are turbulently mixed with additional oxidant and fuel and pass into the burner head and the flame. Large droplets deposit and pass down the drain; 85 to 90% of the sample is discarded in this way.

Flame atomic absorption spectrometry was developed in (ironically) the year 1955 independently by Walsh in Australia and by Alkemade and Milatz in the Netherlands. Because electrons in quantized energy states for atoms of alkali metal elements can easily be raised to excited states (see above), flame emission spectroscopy is a more appropriate instrumental technique, whereas plasma sources are needed for atoms of most other elements. Atomic absorption spectroscopy is unique in that it uses a flame to create the atomic vapor within which the absorption of radiation from a hollow-cathode lamp HCL can occur. This enables the quantitative determination of some 65 elements, provided a line source can be used. The source of light for AA must produce a narrow band of adequate intensity and stability for prolonged periods. An ordinary monochromator is incapable of yielding a band of radiation as narrow as the peak width of an atomic absorption line. HCLs satisfy these criteria. Refer to the following schematic:

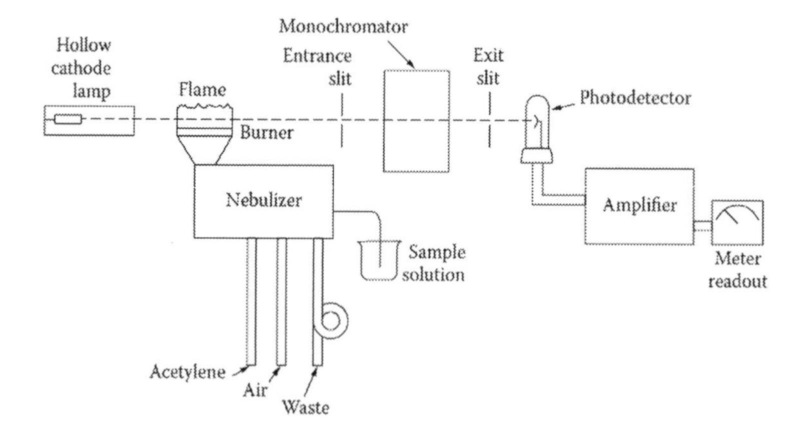

Many models of atomic absorption spectrophotometers are in use in environmental testing laboratories today. Because of this, the type of readouts that one may get might differ. Older instruments most often used *percent absorption*, whereas more contemporary instruments might read out in *absorbance or percent transmittance*. The following schematic relates all three types of AA readouts:

Absorbance	% Transmittance	% Absorption
0	100	0
0.045	90	10
0.097	80	20
0.155	70	30
0.229	60	40
0.301	50	50
0.398	40	60
0.523	30	70
0.699	20	80
1.00	10	90
∞	0	100

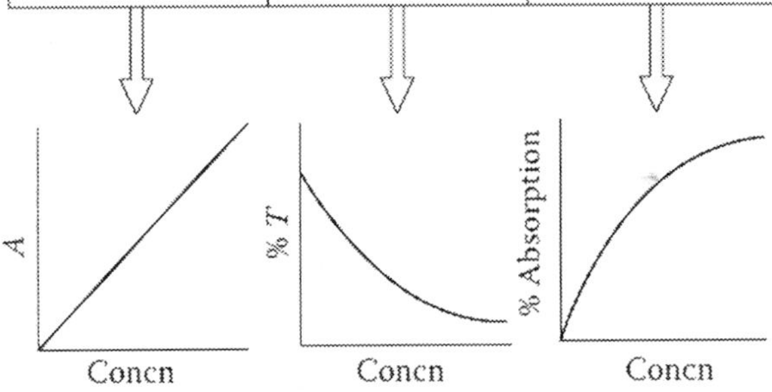

A Concn

% T Concn

% Absorption Concn

This experiment introduces students to quantitative trace metals analysis. The determination of the concentrations of Ca and Mg using FLAA techniques in groundwater is the objective. The PerkinElmer Model 3110® AA needs to be set up in the flame mode of operation. The FLAA will be calibrated and the calibration verified. Various sources of groundwater need to be obtained and used as unknown samples in this experiment.

9.2 EXPERIMENT

9.2.1 PREPARATION OF CHEMICAL REAGENTS

Note: all reagents used in this analytical method contain hazardous chemicals. Wear appropriate eye protection, gloves, and protective attire. The use of concentrated acids and bases should be done in the fume hood.

9.2.2 Chemicals/Reagents Needed per Student or Group

5 mL 1% HNO_3 (use spectroscopic-grade nitric acid only). If unavailable, prepare by placing 3.6 mL of concentrated HNO_3 into a 250 mL volumetric flask previously half-filled with distilled deionized (DDI) water, adjust to the calibration mark with DDI, transfer to a plastic storage bottle, and label "1% HNO_3, spectroscopic grade."
5 mL 1,000 ppm Mg (certified stock solution).
5 mL 1,000 ppm Ca (certified stock solution).

9.2.3 FLAA Operating Analytical Requirement

Specification	Ca	Mg
Optimum range of concentration (minimum #ppm)	0.2	0.02
Optimum range of concentration (maximum #ppm)	7.0	0.05
Wavelength (nm)	422.7	285.2
Sensitivity (ppm)	0.08	0.007
Instrument detection limit (IDL) (ppm)	0.01	0.001

If a wavelength of 202 nm is used for Mg, a wider linear dynamic range is available (i.e. from 0 to 10 ppm). The sensitivity and IDL will be different.

9.2.4 Preparation of the Calibration Curve

As an illustration only, working calibration standards could be prepared as suggested in the following table (this scheme assumes that the 1,000 ppm certified stock reference standard solution has been carefully diluted to 3 ppm).

Standard #	#mL of 3 ppm	Final Volume (mL)[a]	Concentration of Mg (ppm)
0	0	50	0
1	4.15	50	0.25
2	12.5	50	0.75
3	25	50	1.5
ICV	10	50	0.6

[a] Use 1% HNO3 for all dilutions.

9.2.5 Procedure

Detailed instructions on how to set up and operate an FLAA using the Model 3110® AA will be made available in the laboratory. The burner–nebulizer attachment will need to be installed and the energy throughput optimized using either the Ca or Mg

HCL. Proceed to aspirate the working calibration standards. An instrument calibration verification (ICV) standard should be prepared whose concentration should be approximately between the low and high standards. It should be sufficiently aspirated so that triplicate determinations of the absorbance from the same ICV solution can be made. If the precision and accuracy for the calibration and ICV are found acceptable, then any and all available unknowns can be run. A fortified sample containing both Ca and Mg should be available. This sample may have to be diluted if the absorbance is found to significantly exceed the linear dynamic range of the instrument. Be sure to *record the name or the correct code* for each unknown sample. Establish the IDL experimentally for your instrument for each of the two metals. This is accomplished by obtaining seven replicate absorbance measurements of your blank standard.

9.3 FOR THE LAB NOTEBOOK (NO REPORT NECESSARY)

Conduct a determination of the IDL for each metal by measuring the signal due to a blank that is repeatedly aspirated into the flame seven times. Apply the principles of blank calibrations and calculate the IDL assuming $k = 3$. Establish a least-squares regression fit to the experimental calibration data points for both metals. Calculate the standard deviation in both the slope and y-intercept for the calibration curve for both metals. Calculate the standard deviation in the measurement of the ICV for triplicate absorbance measurements. Calculate the confidence limits for the ICV for each metal. Calculate a percent error in the ICV for each metal. Report the concentrations for both Ca and Mg in the fortified sample and for one or more unknown groundwater samples. Comment on the differences in IDLs for Ca and Mg using FLAA.

9.4 SUGGESTED READING

To develop this experiment, the author consulted the following resources:

Methods 7140 (Ca) and 7450 (Mg). In *Test Methods for Evaluating Solid Waste, SW-846*, 3rd edn. Washington, DC: EPA Office of Solid Waste, 1986.

Sawyer D., W. Heinemann, J. Beebe. *Chemistry Experiments for Instrumental Methods*. New York: Wiley, 1984, pp. 242–265.

Skoog D., J. Leary. *Principles of Instrumental Analysis*, 4th edn. Philadelphia, PA: WB Saunders, 1992. Chapter 10 provides one of the better discussions of FLAA. Figure 10.15 provides an excellent drawing of the laminar flow burner for FLAA.

Students who might be interested in finding the original publications that led to the development of the FLAA determinative technique will find citations to the following original literature:

Alkemade C., J Milatz. *J Opt Soc Am* 45: 583, 1955.

Walsh A. *Spectrochimica Acta* 7: 108, 1955.

10 Determination of Lead in Drinking Water Using Graphite Furnace Atomic Absorption Spectroscopy (GFAA)

External Standard vs. Standard Addition Calibration Mode

10.1 BACKGROUND AND SUMMARY OF METHOD

Unfortunately, among the so-called heavy metals that have made their way into the environment, *lead (Pb) is considered extremely toxic and its presence must be identified and its concentration measured* in air, water, and soil. Two principal historical uses in which Pb was contained were in coatings, e.g. house paint as PbCO3 (white) and $PbCrO_4$ (yellow), and in gasoline, e.g. as the organo-lead compound tetraethyl lead. This compound was added to gasoline to boost octane ratings. Older water pipes throughout the US were made of elemental Pb. The extreme toxicity of Pb has required that *instrumental analytical techniques that offer the lowest possible detection limits be used*. The Lead and Copper Rule established by the US federal government on June 7, 1991, states the following: "To protect public health by minimizing lead (Pb) and copper (Cu) levels in drinking water, primarily by reducing water corrosivity. Pb and Cu enter drinking water mainly from corrosion of Pb and Cu containing plumbing materials." The lead and copper rule established *action levels* of 0.015 mg/L for Pb and 1.3 mg/L for Cu in drinking water. Pb joins elements such as Hg, Cd, As, and Tl as requiring determinative techniques with the lowest instrument detection limits (IDLs) possible. It is important to recognize the difference between method detection limits (MDLs) and IDLs. The MDL incorporates the IDL and is equal to the IDL if and only if there is no sample preparation involved. Sample preparation is more common in trace organics analysis in contrast to trace metals analysis. For example,

DOI: 10.1201/9781003260707-10

the IDL for Pb using flame AA is ~0.1 ppm, whereas the IDL for Pb using graphite furnace atomic absorption spectroscopic (GFAA) methods is ~1 ppb.

When using the GFAA technique, a representative aliquot of a sample is placed in the graphite tube in the furnace, evaporated to dryness, charred, and atomized. A greater percentage of available analyte atoms is vaporized and dissociated for absorption in the tube in contrast to FLAA. It becomes possible to use smaller sample volumes. Radiation from a given excited element is passed through the vapor containing ground-state atoms of that element. The intensity of the transmitted radiation decreases in proportion to the amount of the ground-state element in the vapor. The metal atoms to be measured are placed in the beam of radiation, which is nearly monochromatic (from a hollow-cathode tube), by increasing the temperature of the furnace. A monochromator isolates the wavelength of the transmitted radiation and a photosensitive device measures the attenuated intensity. Beer's Law of Spectrophotometry applies, as was the case for UV-vis absorption spectrophotometry, and the concentration of the specific element is determined by various modes of analyte calibration in a manner similar to that for UV absorption spectrophotometry. A schematic of an electrothermal atomizer follows:

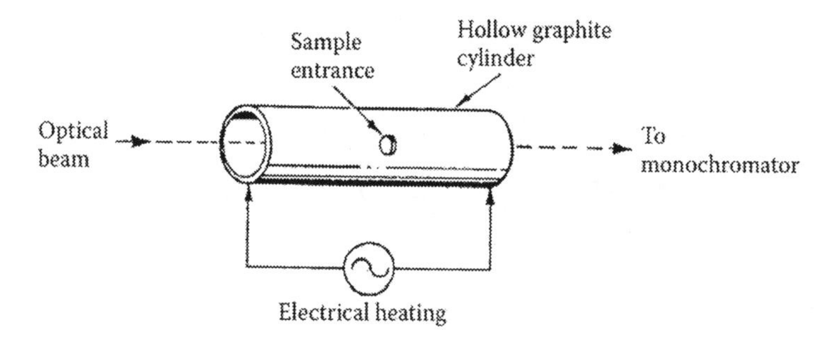

The tube is usually coated with pyrolytic graphite, which is made by heating the tube in a methane atmosphere. Pyrolytic graphite exhibits a low gas permeability and good resistance to chemical attack. This feature lengthens the lifetime (i.e. the number of successful firings) of the tube. There is, however, a finite lifetime for each tube in GFAA.

It is generally believed that the atomization mechanism for any metal M involves reduction of the solid oxide on the graphite surface according to

$$MO_{(s)} + C_{(s)\rightarrow} M_{(l)} \rightarrow M_{(g)} + 0.5M_{2(g)}$$

Most commercial electrothermal atomizers based on the L'vov furnace, as simplified by Massman, undergo vigorous changes in tube temperature. The analyte atoms volatilized from the tube wall come into a cooler gas, so that molecular species that are not detected are formed. This leads to what is termed matrix interferences. For example, equal concentrations of Pb^{2+} in a matrix of deionized water vs. one of high chloride content *would yield different absorbances.* One way to remove this chloride matrix interference is to use a matrix modifier. The addition of an ammonium salt to the chloride-containing matrix would cause volatilization of NH_4Cl with the removal of chloride from the sample matrix.

Each end of the furnace tube is connected to a high-current, programmable power supply through water-cooled contacts. The power supply controlling the furnace can be programmed to dry, ash, and atomize the sample at the appropriate temperatures. The temperature and duration of each of these steps can be controlled over a wide range. The optimization of the operating conditions is very important in developing methods using GFAA. A description of the three major temperature program changes, known as steps, is important in understanding GFAA operation. In the first step (drying), the solvent is evaporated at a temperature just above its boiling point. For aqueous solutions, the temperature is held at 110° C for about 30 sec. In the second step (ashing), the temperature is raised to remove organic matter and as many volatile components from the sample matrix as possible without analyte loss. The ashing temperature varies from 350 to 1,200° C and is maintained for about 45 sec. The last step (atomization) occurs between 2,000 and 3,000° C and lasts for just a few seconds. The element of interest is atomized and the absorption of the source radiation by the atomic vapor is measured. The furnace is then cleaned by heating the atomizer to the maximum temperature for a short period. Finally, the temperature is decreased to room temperature using water coolant and inert gas flow (argon). This process is depicted graphically below. Note that the graph includes the overall transmitted intensity I_t, the furnace temperature T, and the net absorption after background correction. It becomes important to have a means to subtract out this background. The PerkinElmer Model 3110® AA uses a deuterium background correction technique.

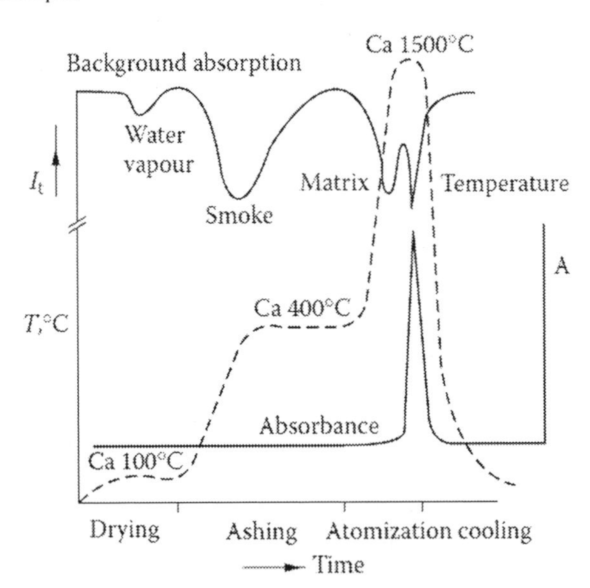

Of the three modes of calibration used in instrumental analysis, external, internal, and standard addition, the latter provides for the *most accurate analytical*

results for samples that exhibit a matrix interference. The external mode of calibration is used to convert the instrumental response to concentration when matrix interferences are not considered a factor. A series of standard solutions that contain the metal of interest are prepared from careful dilution of a *certified standard* stock solution. The calibration curve is obtained and a least-squares regression is performed on the x, y data points. The best-fit line is used to establish the calibration curve. Samples that contain the metal at an unknown concentration level in a sample matrix nearly identical to that used to prepare the serial standards can be run and the data interpolated to give the concentration. In contrast, the standard addition mode of calibration requires that calibration and analysis be performed on the sample itself. Standard addition can be used provided that (1) a linear relationship exists between the physical parameter measured and the concentration of analyte, (2) the sensitivity of the method is not changed by the additions, and (3) the method blank can be corrected for. A typical standard addition calibration follows:

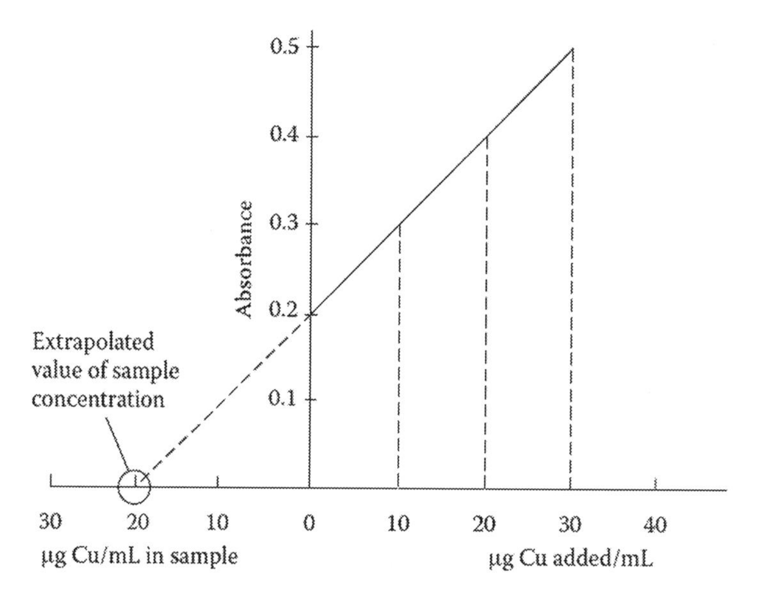

This exercise affords students an opportunity to operate a PerkinElmer Model 3110® GFAA with the objective of determining the concentration of Pb at ppb levels by calibrating the instrument *using two of the three modes: external standard and standard addition.* Following the establishment of the working calibration curve, an instrument calibration verification (ICV) standard containing Pb will be prepared, run in triplicate, and the precision determined. At least one unknown sample will be provided for students to determine the concentration of Pb. *Students are asked to bring in a sample of drinking water to determine the concentration of Pb* once the GFAA has been successfully calibrated and the precision and accuracy for the ICV have been found to be acceptable.

10.2 EXPERIMENT

10.2.1 Preparation of Chemical Reagents

Note: all reagents used in this analytical method contain hazardous chemicals. Wear appropriate eye protection, gloves, and protective attire. The use of concentrated acids and bases should be done in the fume hood.

10.2.2 Reagents Needed per Student or Group

One 40 ppb reference standard containing Pb in 1% spectroscopic-grade nitric acid (HNO_3).
One blank reference standard in 1% spectroscopic-grade HNO_3.
One matrix modifier.
One unknown sample containing Pb. Be sure to record the code.
One drinking water sample from any source.

10.2.3 Procedure

The setup and operation of the GFAA involve the following sequence of activities:

1. Install and align the hollow-cathode lamp (HCL).
2. Align the furnace in the spectrophotometer.
3. Install and condition a new tube.
4. Set up an element parameter file within the PerkinElmer WinLab® software.
5. Align the autosampler.
6. Place blanks, modifiers, and the standard in appropriate locations in the autosampler carousel.
7. Run the calibration according to the programmable sequence within the WinLab® software.
8. Run any and all samples, provided that the calibration and ICV are acceptable.

10.2.4 Using the WinLab® Software

You should find the WinLab® (PerkinElmer) software used to control and acquire GFAA data already downloaded. Retrieve and enter the software via the keyboard. The Pb HCL should turn on as well. A stable signal is necessary in order to continue. Proceed to autozero the detector, and if not satisfied, go to the *realign lamps screen* and adjust the physical position of the Pb HCL so as to bring the energy throughput near to that for the deuterium lamp. Set up a sample sequence in the following order: run a lab blank, then a series of calibration standards for Pb, and then one or more ICVs.

Evaluate the quality of the calibration and, if satisfactory, run the ICV under the *run samples screen*. Retrieve the weight/ID screen and choose an autosampler location and run each sample in triplicate. If precision is acceptable, set up the weight/ID screen with the fortified Pb sample and any other samples of interest.

10.2.5 Preparation of the Stock Reference Pb Standard and Start of the Autosampler

To prepare the calibration standards for Pb, obtain the certified Pb stock standard, which should be 1,000 ppm as Pb. Using an appropriate liquid-handling syringe, take 40 μL of the 1,000 ppm Pb stock and place this aliquot into a 10 mL volumetric flask already half-filled with 1% *spectroscopic-grade HNO3*. Adjust to final volume using the 1% acid. Transfer this solution to a storage container and label "4 ppm Pb." Take 100 μL of the 4 ppm Pb and place this aliquot into a clean 10 mL volumetric flask. Adjust to final volume and transfer to a clean storage container and label "40 ppb Pb." Transfer about a 1 mL aliquot of the 40 ppb Pb reference standard to a plastic GFAA autosampler vial and place it in location 38. Place a clean vial filled with 1% nitric in location 36. Place a clean vial filled with matrix modifier in location 37. The programming for the autosampler is found under *element parameters \ calib*; retrieve this and write the entries into the table in your laboratory notebook.

Using the 40 ppb Pb standard and the WinLab® software, implement an *external mode* of calibration. Retrieve or create a method title for this and conduct the calibration and quantitation. Evaluate whether the calibration is free from systematic error. If so, inject the ICV in triplicate. Record the code for the unknown and inject in triplicate. Run one or more drinking water samples.

Proceed on to the implementation of the *standard addition mode* of calibration using the 40 ppb Pb standard and the WinLab® software. Retrieve the method and conduct the calibration and quantitation. Evaluate whether the calibration is free from systematic error. Inject the ICV in triplicate. Record the code for the unknown and inject it in triplicate. Run one or more drinking water samples. Rerun one or more samples, provided the calibration is linear and the precision and accuracy for the ICV are acceptable.

10.3 FOR THE NOTEBOOK

Include in your notebook data from both modes of calibration. Use an Excel® spreadsheet or other means to calculate the confidence limits at 95% probability using the student's *t* statistics for the ICV for both calibration modes. Report on the concentration of Pb in the coded unknown provided to you. Report on the Pb concentration of any unknown drinking water samples that you analyzed. Proceed to calculate the means, standard deviations, relative standard deviations, and confidence intervals at a given probability. Calculate the relative error between the mean result for the ICV and the expected result for both types of calibration modes (i.e. external standard and standard addition). Comment on the effect of the external vs. standard addition mode of GFAA calibration on the precision and accuracy in the ultra-trace determination of Pb in drinking water.

Discuss why background correction is necessary in AA. Distinguish between physicochemical interferences in AA. Explain what is meant by the term *Smith–Hieftje background correction* and discuss how it differs from the kind of correction employed in the PerkinElmer Model 3110® atomic absorption spectrophotometer.

10.4 SUGGESTED READINGS

Archives of this trade journal are also found. To develop this experiment, the author consulted the following resources:

Massman H. *Spectrochimica Acta B* 23B: 215, 1986.

Method 7000. Atomic absorption methods. In *Test Methods for Evaluating Solid Waste. Physical/Chemical Methods*. SW-846. Revision 1. Washington, DC: Environmental Protection Agency, 1987.

Sawyer D., W. Heinemann, J. Beebe. *Chemistry Experiments for Instrumental Methods*. New York: Wiley, 1984, pp. 242–265.

Skoog D., J. Leary. *Principles of Instrumental Analysis*. 4th edn. Philadelphia, PA: WB Saunders, 1992, pp. 216–218.

Training Manual. Graphite Furnace Atomic Absorption with WinLab® Software. PerkinElmer Inc. Norwalk, CT, 1991.

Vandecasteele C., C. Block. *Modern Methods for trace Elemental Determination*. New York: Wiley, 1993, p. 108.

Yang X. et al. *Am Lab* 51(3): 22–25, 2019. The authors recently reported on using a novel extraction procedure after they derivatized trace concentration levels of the rare Rh3+ metal ion (rhodium) from industrial water processing samples with 2-(5-bromo-2-py ridylazo)-5-dimethylaminoaniline. Rh3+ was quantitated by a GFAA determinative technique. To access this paper, visit the *American Laboratory* online website archives. Archived issues of this trade journal and also found at the Science History Institute located in Philadelphia, PA.

11 A Comparison of Soil Types via a Quantitative Determination of the Chromium Content Using Visible Spectrophotometry and Flame Atomic Absorption Spectroscopy or Inductively Coupled Plasma–Optical Emission Spectrometry

11.1 BACKGROUND AND SUMMARY OF METHOD

Chromium exists in three oxidation states, of which Cr (III) and Cr (VI) are the most stable. Hexavalent Cr is classified as a *known human carcinogen via inhalation*, while Cr(III) is an *essential dietary element* for humans and other animals! Certain soils that exhibit a strong chemically reducing potential have been shown to convert Cr(VI) to Cr(III). It is possible for the analysis of soils from a hazardous waste site to reveal little to no Cr(VI) via the colorimetric method because this method is selective for Cr(VI) only. Because atomic absorption spectrophotometric methods yield total Cr, the *difference between analytical results* from both methods should be indicative of the Cr(III) content of a given soil type. Both methods will be implemented in this laboratory exercise and applied to one or more soil types.

This exercise affords students the opportunity to use several instrumental techniques to which they have previously been introduced in the laboratory in order to

DOI: 10.1201/9781003260707-11

conduct a comparison of soil types with respect to *determining the ratio of Cr(III) to Cr(VI) in terms of their respective concentration* in an environmental soil matrix. The exercise includes pH measurement, calibration of a UV-vis spectrophotometer, calibration of an atomic absorption spectrophotometer in the flame mode (FLAA), and sample preparation techniques.

Cr(VI) in its dichromate form, $Cr_2O_7^{-2}$, reacts selectively with diphenyl carbazide, in acidic media to form a red–violet color of unknown composition. The molecular structure for diphenyl carbazide is shown below:

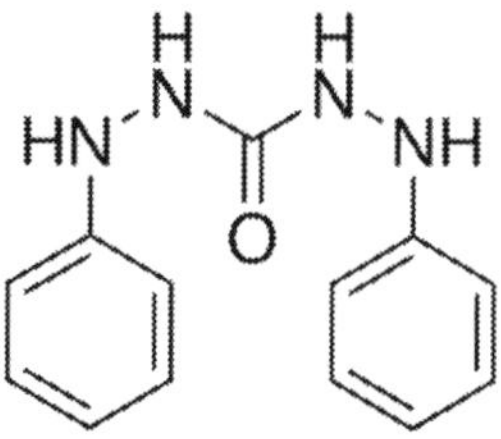

This selectivity for Cr occurs in the absence of interferences such as molybdenum, vanadium, and mercury. The colored complex has a very high molar absorptivity at 540 nm. This gives the method a very low detection limit (MDL) for Cr(VI) using a UV-vis spectrophotometer. Flame atomic absorption spectroscopy is also a very sensitive technique for determining total Cr, with instrument detection limits (IDLs) as low as 3 ppb. Inductively coupled plasma-(atomic) emission spectrometry (ICP-AES) may also be available. Your instructors may choose whether FLAA or ICP-AES is available.

A modification to EPA Method 7196 has been published and will be implemented in this lab exercise (refer to "Suggested Readings"). The method uses a hot alkaline solution (pH 12) to solubilize chromates that are to be found in soils obtained from hazardous waste sites. One portion of the aqueous sample would then be aspirated into the FlAA for a determination of total Cr, whereas diphenyl carbazide dissolved in acetone will be added to another portion, and the absorbance of the red–violet complex will be measured at 540 nm using a visible spectrophotometer. In this manner, both total Cr and Cr(VI) can be determined on the same sample. Thus, the ratio of the concentration of Cr(III) to the concentration of Cr(VI) in a soil sample can be calculated from the data generated in this experiment.

11.2 EXPERIMENT

11.2.1 Chemical Reagents Needed per Student or Group

Note: all reagents used in this analytical method contain hazardous chemicals. Wear appropriate eye protection, gloves, and protective attire. The use of concentrated acids and bases should be done in the fume hood.

Potassium dichromate stock solution: dissolve 141.4 mg of dried $K_2Cr_2O_7$ in distilled deionized water (DDI) and dilute to 1 L (1 mL = 50 μg of Cr).

Potassium dichromate standard solution: dilute 10.00 mL of stock solution to 100 mL (1 mL = 5 μg of Cr).

Sulfuric acid, 10% (v/v): dilute 10 mL of concentrated H_2SO_4 to 100 mL with DDI. Also, 1.8 *M* H_2SO_4 is needed. An aliquot of 10% H_2SO_4 could be used to prepare this solution.

Diphenyl carbazide (DPC) solution: dissolve 250 mg of 1, 5-diphenyl-carbazide in 50 mL of acetone. Store in an amber bottle. Discard when the solution becomes discolored.

Acetone, CH_3COCH_3: use the highest purity available.

Alkaline digestion reagent: 0.28 *M* Na_2CO_3/0.5 *M* NaOH: use your knowledge of chemical stoichiometry to calculate the amount of each base needed to prepare a solution of the desired molarity. Recall that a *1M* solution contains *one mole* of a pure chemical substance per liter of solution. Recall that one mole of a pure chemical substance is its formula or molecular weight in grams.

Concentrated nitric acid, HNO_3.

11.2.2 Procedure for Alkaline Digestion

1. Place 2.5 g of a given soil type into a 250 mL beaker. Add 50 mL of the alkaline digestion reagent. Stir at room temperature for at least 5 min and then heat on a hot plate to maintain the suspensions at 90 to 95° C, with constant stirring for about 1 h. Repeat for all other soil types to be studied whose Cr content is to be determined in this experiment. *Note*: heating on a hot plate can cause bumping and lead to splatter, and thus loss of analyte.
2. Cool the digestates to room temperature, then filter through 0.45 μm cellulosic or polycarbonate membrane filters.
3. Adjust the pH to 7.5 using concentrated HNO_3 and dilute with DDI to a final volume of 100 mL. You now have 100 mL of digestate.

11.2.3 Procedure for Conducting Visible Spectrophotometric Analysis

1. Prepare six working standards from careful dilutions of the 5 μg/mL Cr standard. The range of concentrations should be from 0 to 2 μg/mL Cr. You should prepare 100 mL of each standard.
2. Prepare an initial calibration verification (ICV) standard, which should have its concentration approximately near the mid-range of the calibration. You should prepare 100 mL of the ICV.
3. Prepare a matrix spike and a matrix spike duplicate. The amount of spike should double the concentration found in the original sample. The spike recovery must be between 85 and 115% in order to verify the method.
4. To 45 mL of DDI (this is the method blank), standard, ICV, and digestate, add 1 mL of DPC solution, followed by the addition of 1.8 *M* H_2SO_4 until the pH reaches approximately 2. This should be done in a 125 mL beaker with stirring and immersion of the glass electrode until the desired pH is attained. After cessation of effervescence, dilute the mixture with DDI to

50 mL. Allow the solution to stand from 5 to 10 min. If the solution appears turbid after the addition of DPC, filter through a 0.45 μm membrane. Store the remaining samples and standards in properly labeled bottles. Use if it is necessary to repeat this analysis.

5. Set the spectrophotometer at 540 nm; correctly set the 0 and 100% transmittance settings. Transfer an aliquot of the 50 mL sample to a cuvette. Measure the absorbance of all blanks, standards, and samples. Construct a table in your notebook to facilitate the entry of data.

11.2.4 Procedure for Atomic Absorption Spectrophotometric Analysis or ICP-AES

Refer to SW-846 Methods 7000A ("Atomic Absorption Methods") and 7190 ("Chromium, Atomic Absorption, Direct Aspiration") or Method 6010D ("Inductively Coupled Plasma-Atomic Emission Spectrometry") (ICP-AES) for the quantitative determination of chromium as total Cr. In the lab, proceed to prepare calibration standards and ICVs and aspirate these into the flame using the Model 3110® (PerkinElmer) atomic absorption spectrophotometer. Your instructors may also have the ICP-AES available for you to use such as a Model 2000® (PerkinElmer) ICP-AES. Use the remaining digestates from step 3 and determine total Cr. The FlAA may need to be set up from its present configuration. Refer to previous exercises and training manuals for the necessary information.

11.3 FOR THE REPORT

Include all calibration data, ICVs, and sample unknowns for both instrumental methods. Perform a statistical evaluation in a manner that is similar to that in previous experiments. Use Excel® or an alternative to conduct a least-squares regression analysis of the calibration data. Calculate the accuracy (expressed as a percent relative error for the ICV) and the precision (relative standard deviation for the ICV) from both instrumental methods. Calculate the percent recovery for the matrix spike and matrix spike duplicate. Report on the concentration of Cr in the unknown soil samples. *Be aware of all dilution factors and concentrations as you perform calculations.*

Find the ratio of the concentration of Cr (III) to that of Cr (VI) in each of the soil samples analyzed and present this value at the end of your report.

11.4 SUGGESTED READING

To develop this experiment, the author consulted the following resources:

Budde W. *Mass Spectrometry: Strategies for Environmental and Related Applications.* Washington, DC: American Chemical Society, Oxford University Press, 2001, Chapter 7, pp. 328–347. Provides an excellent introduction to trace environmental elemental analysis.

Method 6010D inductively coupled plasma optical emission spectrometry. In *Test Methods for Evaluating Solid Waste, Physical/Chemical. SW-846. Revision 5*. Washington, DC: Environmental Protection Agency, July 2018.

Method 7000B. Flame atomic absorption spectrophotometry. In *Test Methods for Evaluating Solid Waste, Physical/Chemical. SW-846. Revision 2*. Washington, DC: Environmental Protection Agency, 2007.

Method 7196A. Chromium. Hexavalent (colorimetric). In *Test Methods for Evaluating Solid Waste Physical/Chemical Methods. SW-846, Revision 1*. Washington, DC. Environmental Protection Agency, July 1992.

Skoog D., J. Leary. *Principles of Instrumental Analysis*, 4th edn. Philadelphia, PA: WB Saunders, 1992, Chapter 11, pp. 233–251. This textbook and certainly subsequent editions provide the student with an excellent introduction to the principles of ICP-AES.

Standard Methods for the Examination of Water and Wastewater, 16th edn. Washington, DC: Association of Public Health Association, 1988, pp. 201–204.

Vitale R. et al. *Am Environ Lab* 7:1, 8–10, 1995. This paper discusses a modification to SW-846 Method 7196.

12 Data Acquisition and Instrument Control Using the Turbochrom Chromatography Software: An Introduction to High-Performance Liquid Chromatography (HPLC)

Evaluating Those Experimental Parameters That Influence Separations

12.1 BACKGROUND AND SUMMARY OF METHOD

Contemporary analytical instrumentation is said to be *interfaced to computers*. These developments commenced in the early to mid-1980s and took hold with Microsoft Windows®-based software environments in the 1990s. The architecture for this technological advance can be illustrated as follows:

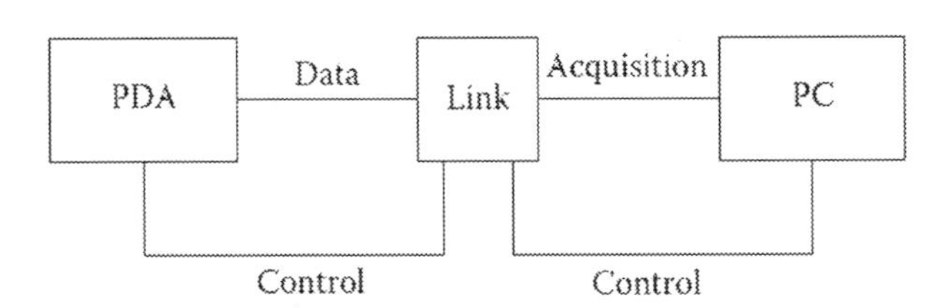

DOI: 10.1201/9781003260707-12

Interfaces can be either stand-alone or installed into the console of the PC. Instruments can be controlled and data acquired from a PC, or if a control is not available, only data acquisition is obtained. In our laboratory, both types of interfaces are used. With appropriate software, the control and data acquisition tasks are easily performed. If a means can be acquired to enable automatic sampling to be controlled as well, a totally automated system can be achieved! This was accomplished in our laboratory.

The HPLC within each student workstation is PC controlled, and the photodiode array detector (PDA) is interfaced to the same PC, thus enabling real-time data acquisition. Students are first asked to study the present architecture so as to gain an appreciation of contemporary HPLC-PDA-DS (data system) technology.

This experiment is designed to take students through an *initial hands-on experience* with the HPLC-PDA-DS from a first sample injection to a simple quantitative analysis. A quick method is first necessary for the software to recognize something. This is followed by optimizing the initial method, conducting a calibration, creating a customized report format, and evaluating the initial calibration verification standard (ICV).

Following completion of the initial experiment, the focus shifts to the separation of a test mixture that contains two organic compounds using the HPLC instrument. The effect of solvent strength on k' and the effect of mobile-phase flow rate on R_S will be considered by retrieving previously developed Turbochrom® methods and making manual injections.

12.1.1 HPLC AND TEQA

High-performance liquid chromatography (HPLC) followed GC in the early development of instrumental column chromatographic techniques that could be applied to TEQA. HPLC almost always complements and, depending on the analyte, sometimes duplicates GC. For example, *polycyclic aromatic hydrocarbons* (PAHs) can be separated and quantitated by both techniques; however, *N-methyl carbamate pesticides* can be determined only by HPLC as a result of the thermal instability in a hot GC injection port. Molecular structures for one example of each organic functional group are shown:

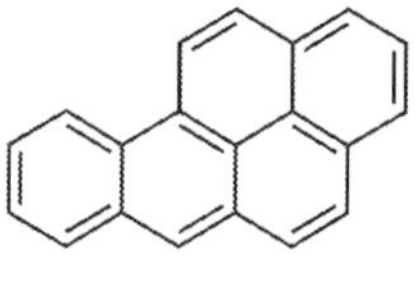

PAH molecule

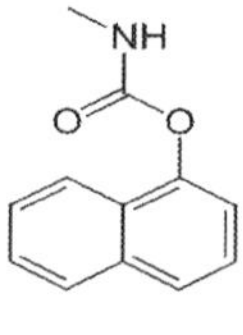

N-methyl carbamate (pesticide Sevin®) molecule

HPLC has become the dominant determinative technique for biochemists and pharmaceutical and medicinal chemists, yet has continued to take a secondary role with environmental chemists until technological advances led to the establishment, initially of HPLC-UV and HPLC-FL, followed by HPLC-PDA during the 1970s through to the 1990s. LC-MS first appeared followed by the LC-MS-MS technique in and around the early 2000s. Samples that contain the more polar and thermally labile analytes are much more amenable to analysis by HPLC rather than by GC. For example, a major contaminant in a lake in California went undetected until State Department of Health chemists identified a sulfonated anionic surfactant as the chief cause of the pollution. This pollutant was found using HPLC determinative techniques. HPLC encompasses a much broader range of applicability in terms of solute polarity and molecular weight range when compared with GC.

To illustrate how these different kinds of HPLC determinative techniques might aid the analyst in the environmental testing laboratory, consider the request from an engineering firm that wishes to evaluate the degree of *phthalate ester contamination* from leachate emanating from a hazardous waste site. Reversed-phase HPLC is an appropriate choice for the separation of lower-molecular-weight phthalate esters (e.g. dimethyl from diethyl from dibutyl). Attempts to elute higher MW and much more hydrophobic (lipophilic) phthalate esters, e.g. dioctyl and bis (2-ethyl hexyl) under reversed-phase conditions were unsuccessful. The separation of these higher MW PAHs under normal-phase HPLC conditions was successful.

12.1.2 Flow-Through Packed Columns

High-performance liquid chromatography requires that liquid be pumped across a packed bed within a tubular configuration. Snyder and Kirkland in their classic text on HPLC have used the Hagen–Poiseuille equation for laminar flow through tubes and Darcy's law for fluid flow through packed beds and derived the following relationship:

$$t_0 = \frac{15{,}000 L^2 \eta}{\Delta P d_p^2 f}$$

where t_0 is the retention time of an unretained solute (the time it takes after injection for an unretained solute to pass through the column and reach the detector), L is the length of the column, η is the viscosity of the mobile phase, ΔP is the pressure drop across the column, d_p is the particle size of the stationary-phase packing, and f is an integer and is 1 for irregular porous, 2 for spherical porous, and 4 for pellicular packings.

The importance of stationary-phase particle size is reflected in the dependence of the void retention volume $V_0 = F \cdot t_0$ where F is the mobile-phase flow rate in #cm^3/min (recall that 1 mL = 1 cm^3) on the *inverse square* of d_p. Recall that the retention volume of a retained solute whose capacity factor is given by k' is

$$V_R = V_0 \left(1 + k'\right)$$

Hence, the smaller the d_p, the larger is V_0 and, consequently, V_R. A smaller d_p also contributes in a significant manner to a larger N.

12.1.3 HPLC Also Refers to an Instrument That Is a High-Pressure Liquid Chromatograph

It is quite useful to view the instrumentation for HPLC in terms of zones according to the following schematic:

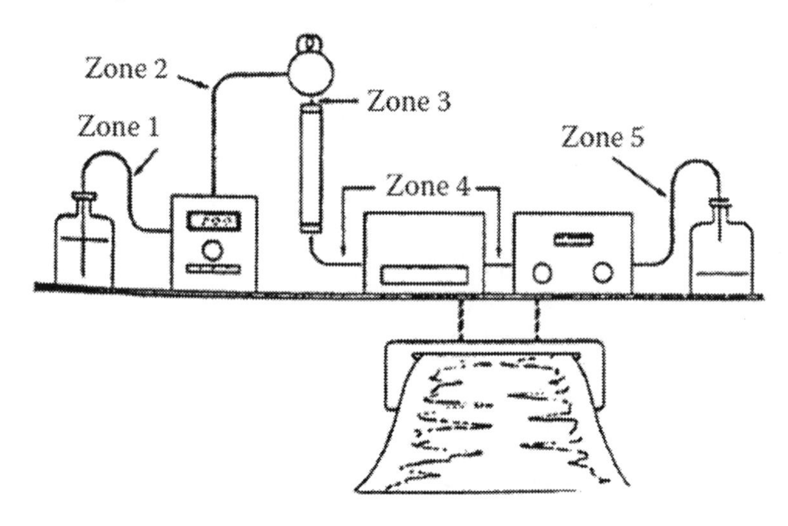

Zone 1—Low-pressure zone prior to the pump. This is a *noncritical area* served by Teflon tubing. A fritted filter is placed at the inlet to prevent particulates from entering the column.

Zone 2—High-pressure zone between the pump and injector. This is a *noncritical area* served by standard stainless-steel (SS) tubing usually 1/16 inch in outer diameter (o.d.). A high-surface-area 0.5 μm filter can be placed here to prevent particulates from reaching the column.

Zone 3—High-pressure area surrounding injector and column. This is a *critical area* where the sample is introduced to the separation system. The volume must be well swept and minimized. Special fittings are 0.25 mm-inner diameter (id) SS tubing.

Zone 4—Low-pressure area between column and detector. In this *critical area*, separation achieved in the column can be lost prior to detection. The volume must be well swept and minimized. Special fittings and 0.25 mm SS or plastic tubing are required. The critical zone extends to all detectors or fraction collectors in series or parallel connections.

Zone 5—Low-pressure area leading to waste collector. This *noncritical area* is served by Teflon tubing. Most labs fail to fit the waste vessel with a vent line to the hood or exhaust area.

12.2 EXPERIMENT

High-performance liquid chromatograph (HPLC) instrument incorporating a UV absorption photodiode array detector (PDA) under reversed-phase liquid chromatographic conditions.

12.2.1 Preparation of Chemical Reagents

Note: all reagents used in this analytical method contain hazardous chemicals. Wear appropriate eye protection, gloves, and protective attire. The use of concentrated acids and bases should be done in the fume hood.

12.2.2 Accessories to Be Used with the HPLC per Group

One HPLC syringe. This syringe incorporates a blunt end; the use of a beveled-end GC syringe would damage inner seals to the Rheodyne injector.

One 10 mL two-component mix at 1,000 ppm each. Prepare the mixture by dissolving 10 mg of ortho-phthalic acid (PhtA) and 10 mg of dimethyl ortho-phthalate (DMP) in about 5 mL of 50:50 (ACN: H_2O) in a 50 mL beaker. After dissolution, transfer to a 10 mL volumetric flask and adjust to the final mark with the 50:50 solution. Molecular structures for PhtA and DMP are shown here:

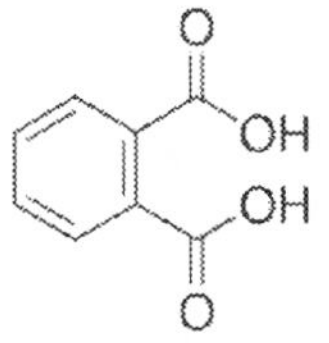

ortho-phthalic acid (PhtA)

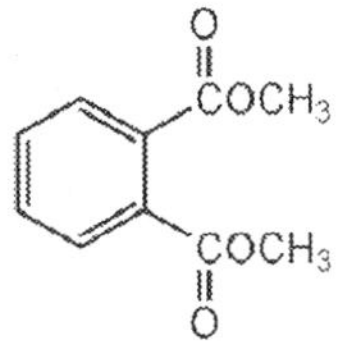

dimethyl ortho-phthalate (DMP)

12.2.3 Procedure

Be sure to record your observations in your laboratory notebook.

12.2.3.1 Initial Observations of a Computer-Controlled High-Performance Liquid Chromatograph

Upon approaching the HPLC-PDA-DS, conduct the following:

1. Identify each of the five zones discussed above.
2. Locate the following hardware components:
   a. The IEEE-488 cable to the LINK interface.
   b. The start/stop line from the Rheodyne injector to the LINK.
   c. The data acquisition line from the PDA to the LINK.
   d. The keying and master key.

12.2.3.2 Creating a QuickStart Method, Acquiring Data, Optimizing, Calibrating, and Conducting Analysis Using the QuickStart Method

Proceed with the Turbochrom 4 Tutorial and create a method using QuickStart. Inject an aliquot of a 100 ppm test mix reference standard. Optimize the method using the Graphic Editor. Develop the calibration and report format sections of your method.

Establish a three-point calibration for DMP only between 10 and 100 ppm (always inject reference standards from low concentration to high concentration, never the reverse!) and prepare an ICV. Run the ICV in triplicate.

12.2.3.3 Effect of Solvent Strength on k'

A good practice when beginning to use a RP-HPLC instrument is to initially pass a mobile phase that contains 100% acetonitrile (ACN) so as to flush out of the reversed-phase column any nonpolar residue that might have been retained from previously running the instrument. Retrieve the Turbo method titled *100% ACN* and *download* if not already set up. *Download within "setup" using the "method" approach.*

Retrieve the method from *Turbochrom* or equivalent software titled *80% ACN* and proceed to use *Setup in the method mode* to enable you to operate the HPLC with a mobile-phase composition of 80% ACN and 20% aqueous. The use of *Setup* is called *downloading* the method and sequence file so that data acquisition can begin. The aqueous mobile phase consists of 0.05% phosphoric acid (H_3PO_4) dissolved in distilled deionized water (DDI). Carefully fill the 5 μL injection loop (the injector arm should be in the "load" position with the evaluation test mix with the HPLC syringe). Inject by moving the injector arm from the "load" position to the "inject" position. *Observe the chromatogram that results and note the retention times of the components in the mixture.* Give all members in the group the opportunity to make this initial injection.

Retrieve the method titled *60%ACN, download it*, and then proceed to repeat the injection procedure discussed above. *Observe the chromatogram that results and note retention times.*

Retrieve the method titled *40%ACN, download it*, and then proceed to repeat the injection discussed earlier. *Observe the chromatogram that results and note retention times.*

Retrieve the method titled *20%ACN, download it*, and then proceed to repeat the injection procedure discussed earlier. *Observe the chromatogram that results and note retention times.*

12.2.3.4 Effect of Mobile-Phase Flow Rate on Resolution

The mobile-phase flow rate will be varied and its influence on chromatographic resolution will be evaluated.

Retrieve the method titled *FlowHi* and *download the method* and then proceed to use *Setup* as you did during the variation of solvent strength experiments. Allow sufficient equilibration time at this elevated mobile-phase flow rate. Notice what happens to the column back-pressure when a higher flow rate is in operation. Inject the test mix and *observe the chromatogram that results.*

Retrieve the method titled FlowLo and *download the method*, and then proceed to repeat the injection procedure discussed earlier. *Observe the chromatogram that results.*

12.3 FOR THE LAB NOTEBOOK

The following empirical relationship has been developed for RP-HPLC. Refer to a theoretical discussion on HPLC or to a more specialized monograph.

$$\log k' = \log k_W - S\Phi$$

where *k;* is the capacity factor for a retained peak, k_W is the capacity factor (extrapolated) *k'* for pure water, Φ is the volume fraction of the organic solvent in the mobile phase, and *S* is a constant that is approximately proportional to solute molecular size or surface area.

Choose one component in the evaluation test mix and determine whether the above equation is consistent with your observations.

Address the following:

1. Among the three major parameters upon which resolution R_S depends, which of the three is influenced by changes in mobile-phase flow rate? Explain.
2. Mr. Everett Efficient believes that he can conserve resources by operating his HPLC using a mobile phase that consists only of a 0.01 *M* aqueous solution containing sodium dihydrogen phosphate (NaH_2PO_4). Discuss what is seriously deficient in Mr. Efficient's fundamental assumption.
3. Assume that you could change HPLC columns in this exercise and that you installed a column that contained *3 μm particle size* silica. Assume that you used the same mobile-phase composition that you used for the reversed-phase separations that you observed. Explain what you would expect to find if the reversed-phase test mix were injected into this HPLC configuration.
4. Explain why DMP is retained longer (i.e. has the higher *k')* than PhtA given the same mobile-phase composition.

12.4 SUGGESTED READING

To develop this experiment, the author consulted the following resources:

Ahuja S. *Selectivity and Detectability in HPLC. Chemical Analysis Series of Monographs on Analytical Chemistry and Its Application*. Vol. 104. New York: Wiley Interscience, 1989, p. 28.

Guide to LC. Woburn, MA: Ranin Instruments Corporation, n.d.

Sawyer D., W. Heineman, J. Beebe. *Chemistry Experiments for Instrumental Methods*. New York: Wiley, 1984, pp. 344–360.

Snyder L., J. Kirkland. *Introduction to Modern Liquid Chromatography*. 2nd edn. New York: Wiley, 1979, pp. 36–37.

Snyder L., J. Kirkland, J. Glajch. *Practical HPLC Method Development*. 2nd edn. New York: Wiley, 1997.

13 Identifying the Ubiquitous Phthalate Esters in the Environment Using HPLC, Photodiode Array Detection, and Confirmation by GC-MS

13.1 BACKGROUND AND SUMMARY OF METHOD

The most commonly found organic contaminant in landfills and hazardous waste sites has proved to be the homologous series of aliphatic esters of phthalic acid. This author has *found phthalate esters in almost every Superfund waste site sample GC-MS report* that he reviewed during the mid-1980s while consulting for an environmental testing laboratory in New York State!

The molecular structures for two representative phthalate esters are drawn in the following figure. Dimethyl phthalate (DMP) and bis (2-ethylhexyl) phthalate (Bis) illustrate one example of a lower-molecular-weight phthalate ester versus a higher-molecular-weight ester. DMP and the higher homologs, diethyl phthalate (DEP), di-*n*-propyl (DPP), and di-*n*-phthalate (DBP), are the focus of this exercise.

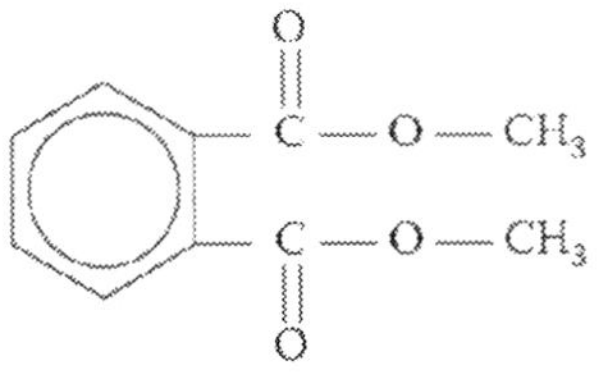

$COOCH_2\,CH(C_2H_5)(CH_2)_3\,CH_3$

$COOCH_2\,CH(C_2H_5)(CH_2)_3\,CH_3$

DMP 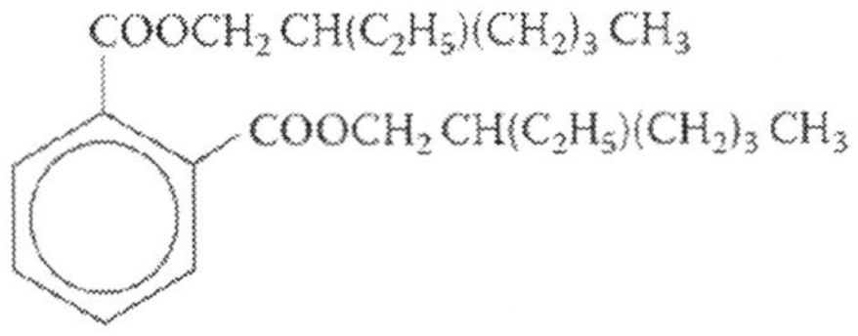 Bis

The photodiode array UV absorption detector provides both spectral peak matching and, if desired, peak purity determinations. This is nicely illustrated below. A peak is identified in the *first drawing* and its UV absorption spectrum can be matched against a library of UV absorption spectra. Note that the UV absorption spectrum from the peak at or near a retention time t_R = 39 min in the HPLC chromatogram is retrieved from a stored library file. The UV spectrum for the peak and that for a

DOI: 10.1201/9781003260707-13

reference standard are compared. The *second drawing* demonstrates how overlays of UV absorption spectra use three points across the chromatographically resolved HPLC and are used together with an algorithm to calculate a purity match. Note the difference between the overlaid UV absorption spectra for the impure vs. the pure peak. You will not be using the peak purity algorithm in this exercise.

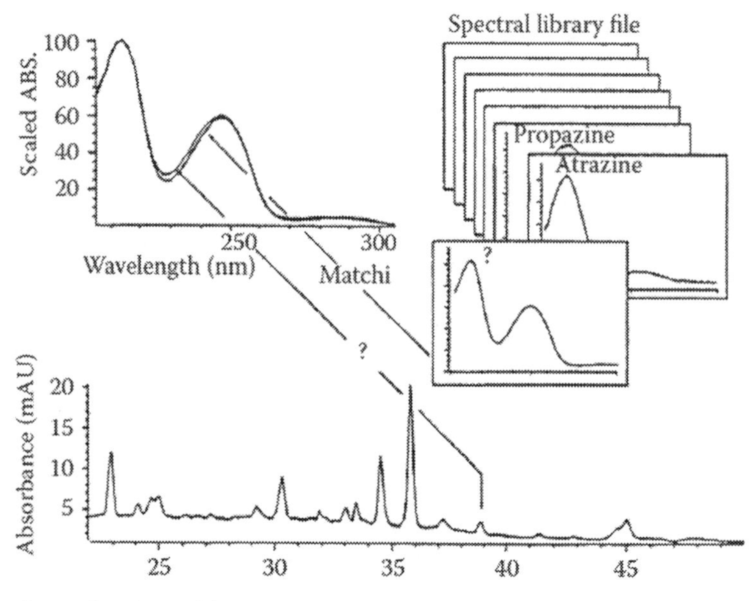

Spectral peak matching.

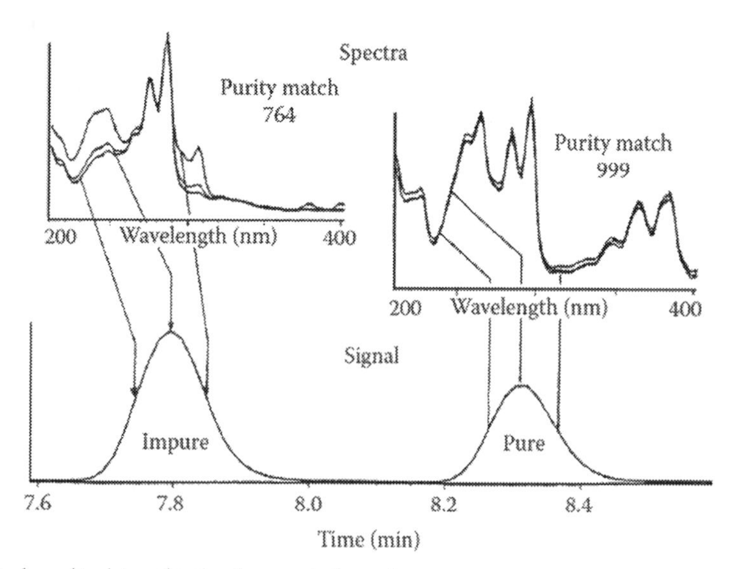

Peak purity determination by spectral overlay.

13.1.1 Analytical Method Development Using HPLC

Analytical method development in HPLC usually involves changing the composition of the mobile phase until the desired degree of separation of the targeted organic compounds has been achieved. One starts with a mobile phase that has a *high solvent strength* and moves downward in solvent strength to where a satisfactory resolution can be achieved. Recall the key relationship for chromatographic resolution:

$$R_s = \frac{1}{4}(\alpha - 1)(N)^{\frac{1}{2}}\left(\frac{k'}{1+k'}\right)$$

A useful illustration of the effects of selectivity, plate count, and capacity factor is shown below:

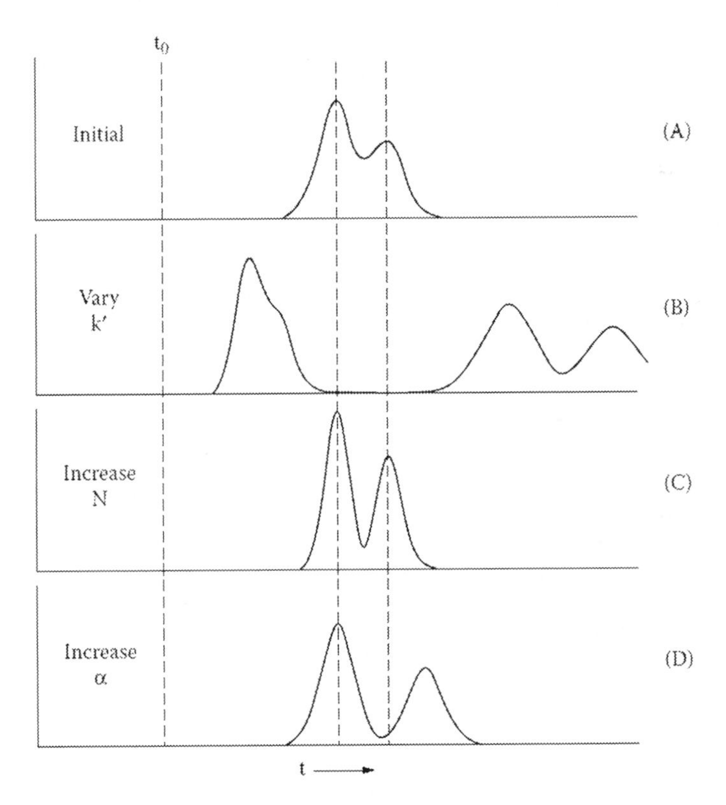

HPLC chromatogram (A) shows a partial separation of two organic compounds, e.g. DMP from DEP. This degree of resolution, R_S, could be improved by changing k', N, or α. In (B), k' is increased, which changes the retention times and shows a slight improvement in R_S. Increasing N significantly increases R_S, as shown in (C); the greatest increase in R_S is obtained by increasing α, as shown in (D). Refer to the "Suggested Reading" at the end of this experiment or an appropriate textbook chapter or monograph on HPLC to enlarge on these concepts.

13.1.2 GC-MS Using a Quadrupole Mass Spectrometer

In a manner similar to obtaining specific UV absorption spectra for chromatographically separated peaks, as in HPLC-PDA, GC-MS also provides important identification of organic compounds first separated by gas chromatography. The mass spectrometer that you will use consists of four rods arranged to form parallel sides of a rectangle, as shown below. The beam from the ion source is directed through the quadrupole section, as shown below.

The quadrupole rods are excited with a large DC voltage superimposed on a radio frequency (RF) voltage. This creates a three-dimensional, time-varying field in the quadrupole. An ion traveling through this field *follows an oscillatory path*. By controlling the ratio of RF to DC voltage, ions are selected according to their mass-to-charge ratio (*m/z*). Continuously sweeping the RF/DC ratio will bring different *m/z* ratios across the detector. An oversimplified sketch of a single quadrupole MS appears below:

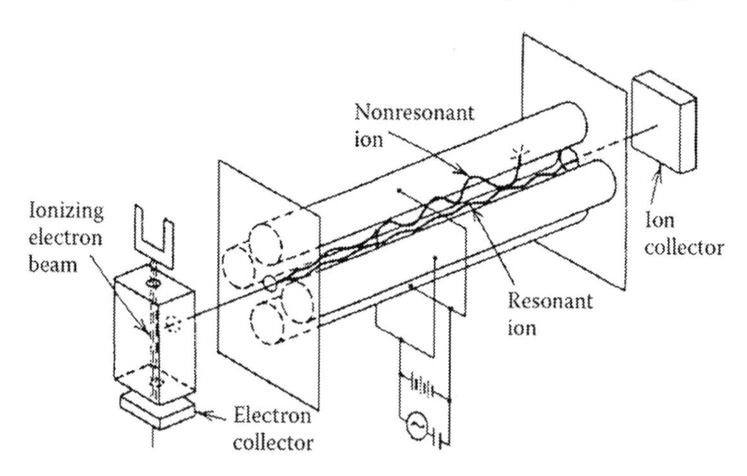

13.2 OF WHAT VALUE IS THIS EXPERIMENT?

The goal of this experiment is to provide an opportunity for students to engage in analytical method development by *identifying an unknown phthalate ester* provided to them. This is an example of *qualitative analysis*. The reference standard solution consists of a mixture of the four phthalate esters: dimethylphthalate (DMP), diethylphthalate (DEP), di-n-propylphthalate (DnPP), and di-n-butylphthalate (DnBP). Molecular structures for these four *phthalate esters* are shown here:

DMP DEP DnPP DnBP

Each group will be given an unknown that contains one or more of these phthalate esters. A major objective would be to use available instrumentation to identify the unknown phthalate ester! Students will have available to them an HPLC in the reversed-phase mode (RP-HPLC) and also access to the department's gas chromatograph-mass spectrometer (GC-MS) system.

Students must first optimize the separation of the esters using RP-HPLC, record and store the ultraviolet absorption spectra of the separated esters, and compare the spectrum of the unknown against the stored UV spectra. In addition, staff will be available to conduct the necessary GC-MS determination of the unknown. A hard copy of the chromatogram and mass spectrum will be provided so that the student will have additional confirmatory data from which to make a successful identification of the *unknown phthalate ester*.

13.3 EXPERIMENT

A high-performance liquid chromatograph with ultraviolet absorption photodiode array detection (HPLC-PDA) is set up for reversed-phase liquid chromatographic separations.

A capillary gas chromatograph-mass spectrometer incorporating a quadrupole mass-selective detector (C-GC-MS) is used. This instrument should be available for students to use or to drop off their samples at a location outside of the instrumental teaching laboratory location.

13.3.1 Preparation of Chemical Reagents

Note: all reagents used in this analytical method contain hazardous chemicals. Wear appropriate eye protection, gloves, and protective attire. The use of concentrated acids and bases should be done in the fume hood.

13.3.2 Accessories to be Used With the HPLC per Student or Group

One HPLC syringe. This syringe incorporates a blunt end; the use of a beveled-end GC syringe would damage inner seals to the Rheodyne HPLC injector.

One four-component phthalate ester standard. Check the label for concentration values.

One unknown sample that contains one or more phthalate esters. Be sure to record the code for the unknown assigned.

13.3.3 Procedure

Unlike other lab exercises, *no methods have been developed for this exercise*. Consult with your lab instructor regarding the details for developing a general strategy. You will be introduced to Turboscan®, software that will allow you to store and retrieve UV absorption spectra.

First, find the mobile-phase solvent strength that optimizes the separation of the four phthalate esters. *Second*, retrieve the UV absorption spectrum for each of the four and build a library. *Third*, inject the unknown sample and retrieve its UV spectrum. *Fourth*, make arrangements with the staff to get your unknown sample analyzed using GC-MS.

13.4 FOR THE REPORT

Include your unknown phthalate ester identification code along with the necessary laboratory data and interpretation of results to support your conclusions.

Please address the following in your report:

1. Compare the similarities and differences for the homologous series of phthalate esters on both UV absorption spectra and mass spectra from your data.
2. Explain what you would have to do if you achieved the optimum resolution and suddenly ran out of acetonitrile! Assume that you have only methanol available in the lab. Would you use the same mobile-phase composition in this case?
3. This exercise introduces you to the quadrupole mass filter. Briefly describe how the mass spectrum is obtained, and if you so desire, attempt to provide a brief mass spectral interpretation. You may want to review a text that discusses the principles and practice of GC-MS.

13.5 SUGGESTED READING

To develop this experiment, the author consulted the following resources:

Sawyer D., W. Heineman, J. Beebe. *Chemistry Experiments for Instrumental Methods*. New York: Wiley, 1984, pp. 344–360.

Snyder L., J. Kirkland. *Introduction to Modern Liquid Chromatography*. 2nd edn. New York: Wiley, 1979.

Snyder L., J. Glajch, J. Kirkland. *Practical HPLC Method Development*. New York: Wiley, 1988.

Snyder L., J. Kirkland, J. Glajch. *Practical HPLC Method Development*. 2nd edn. New York: Wiley, 1997.

To see how the LC-MS-MS determinative technique has significantly impacted the practice of TEQA, refer to two recent contributions:

Parry E., T. Anumol. Quantitative analysis of perfluoroalkyl substances (PFAS) in drinking water using liquid chromatography-tandem mass spectrometry. *Current Trends in Mass Spectrometry*, October 2019, pp. 21–24.

Shrestha P. et al. *Current Trends in Mass Spectrometry*, October 2019, pp. 7–15. The authors compared LC-MS-MS quantitation results obtained between APPI vs. ESI interface with respect to wastewater analysis for various pharmaceuticals.

Current Trends in Mass Spectrometry is published (as of this writing) by MultiMedia Healthcare LLC which also publishes monthly issues of *LC-GC North America*, *LC-GC Europe*, and *Spectroscopy*. Interested readers who qualify can receive the print and e-versions without cost. Issue archives are also available online.

14 An Introduction to Gas Chromatography

Evaluating Experimental Parameters That Influence Gas Chromatographic Performance

14.1 BACKGROUND AND SUMMARY OF METHOD

Gas chromatography (GC) is *the* most widely used instrumental technique for the determination of trace concentrations of volatile (VOC) and semi-volatile (SVOC) organic pollutants found in environmental samples today! Its origins stem from the pioneering work of Martin and Synge in 1941 to the development of open tubular gas chromatographic columns advanced by Golay to the fabrication by Dandeneau and Zerenner (*J High Resol. Chromat. Chromatgr. Commun 2 351[1979]*) at Hewlett-Packard of the *fused-silica* wall-coated open tubular (WCOT) gas chromatographic column which revolutionized the practice of gas chromatography. It must be recognized, however, that approximately 20% of all of the organic compounds that exist and could possibly make their way into the environment are amenable to GC techniques without prior chemical modification! Despite this limitation, over 60 organic compounds classified as VOCs have been found in drinking water, groundwater, surface water, and wastewater and are routinely monitored. Over 100 SVOCs have also been found, which include phenols, polycyclic aromatic hydrocarbons, mono-, di-, and trichloro aromatics and aliphatics, nitroaromatics, polychlorinated biphenyls, organochlorine pesticides, organophosphorus pesticides, triazine herbicides, and phthalate esters, among others.

The theoretical principles that underlie GC are found in numerous texts and monographs. Specific methods that incorporate GC as the determinative instrumental technique are to be found in a plethora of analytical methods published by the Environmental Protection Agency (EPA), the American Public Health Association/American Water Works Association/Water Pollution Control Federation (*Standard Methods for the Examination of Water and Wastewater*), and the American Society for Testing Materials (ASTM, Part 31, *Water*).

DOI: 10.1201/9781003260707-14

This exercise introduces the student to those experimental GC parameters that exert a major influence on GC performance. These include (1) detector selectivity, (2) injection volume vs. chromatographic peak shape, (3) the effect of changing the carrier gas flow rate on column efficiency, and (4) the effect of column temperature on chromatographic resolution and analysis time. This exercise affords the student the opportunity to vary these parameters and assess the outcomes. *This is a qualitative analysis exercise only and involves making and recording observations and doing some calculations from information found in the chromatograms.*

14.2 BRIEF DESCRIPTION OF GAS CHROMATOGRAPHS LOCATED IN THE HAZARDOUS WASTE ANALYSIS LAB AT MICHIGAN STATE UNIVERSITY

The Autosystem® PerkinElmer gas chromatograph (GC) consists of a dual-injector port, dual-capillary-column configuration, and dual detectors including the flame ionization detector (FID) and the electron-capture (ECD) connected via an analog-to-digital (A/D) interface to a personal computer (PC) workstation. The PC is driven by the Turbochrom® (PE Nelson) chromatography data processing software. You will encounter two types of A/D interfaces in the laboratory. The 600 LINK interface provides for both data acquisition and instrument control. The 900 interface provides for data acquisition only.

The *front injector* consists of a split/splitless capillary column type and is connected to a 0.25 mm (i.d.) × 30 m (length) wall-coated open tubular (WCOT) column (referred to as a narrow-bore WCOT column). The column is coated with a DB-5 liquid phase (5% phenyl dimethyl siloxane) that is chemically bonded to the surface within the WCOT. This type of liquid phase is appropriate for the separation of SVOCs whose boiling points are much greater than 100° C. The optimum volumetric flow rate (i.e. the flow rate that gives a minimum in the *van Deemter* curve) is between 1 and 3 *cm^3 per minute* (cm^3/min) To obtain such a low flow rate, a split vent is required to remove most of the gas. Refer to the instruction manual for *setting the split ratio.* Typical split ratios are 1:25, 1:50, or 1:100, and this ratio refers to the ratio of gas flow through the column to that through the vent. The outlet end of this column is connected to the inlet to the ECD. This detector requires an additional source of inert gas, commonly called *makeup gas.* The flow rate for the makeup should be approximately 30 cm^3/min. Using the digital flow check meter (refer to the instruction manual), measure the initial flow rate, then adjust to the optimum for the operation of a narrow-bore WCOT column.

The *rear injector* consists of a *packed column adapted* for connection to a 0.53 mm (i.d.) × 30 m (length) capillary column (referred to as a *megabore* column). The column is coated with a cyanopropyl dimethyl polysiloxane liquid phase that is chemically bonded to the inner tubing wall. This type of liquid phase is appropriate for the separation of VOCs. The optimum volumetric flow rate is between 5 and 15 cm^3/min. The column outlet is connected to the inlet to the FID. This detector does not require makeup gas. The FID, however, requires a 10:1 ratio of airflow to hydrogen flow. Conventional flow rates are 300 cm^3/min for air and 30 cm^3/min for H_2. Once the air/fuel ratio has been established, the FID can be ignited. Sometimes, a slightly fuel-rich ratio is necessary to ignite the FID.

14.3 PRINCIPLE OF SEPARATION IN GC

When two compounds migrate at the same rate through a chromatographic column, no separation is possible. Two compounds that differ in retention times, abbreviated t_R, or capacity factor, abbreviated k' and appear to separate, do so because of differences in their equilibrium distribution constants, denoted by K_D. If K_D is independent of sample size, Gaussian elution bands (i.e. symmetrical peaks) are observed. This is the case with *linear elution chromatography*. In other words, a plot of the concentration of analyte in the stationary phase to the concentration of analyte in the mobile phase yields a straight line whose slope equals K_D. If the amount of analyte increases either by injecting equal volumes of solutions whose concentrations are increasing or by injecting increasing volumes of a solution whose concentration is fixed, nonsymmetrical chromatographic peaks result. K_D is now dependent on the amount of solute, and either peak tailing or peak fronting results. This is the case of nonlinear elution chromatography. Gaussian or symmetrical peak shape is a chief objective when GC is used to perform trace quantitative analysis. The following equations relate to the parameters discussed above:

$$k' = \frac{t_R - t_0}{t_0}$$

$$K_D = \beta k'$$

where β is the ratio of the volume of the mobile phase to the volume of the stationary phase; t_R is the retention time for a retained peak, while t_0 is the retention time for an un-retained peak.

14.4 EXPERIMENT

Gas chromatograph interfaced to a PC that is loaded with chromatographic software. In our lab, an Autosystem® (PerkinElmer) is interfaced with a PC workstation that utilizes Turbochrom® (PE Nelson) for data acquisition, processing, and readout.

14.4.1 Preparation of Chemical Reagents

Note: all reagents used in this analytical method contain hazardous chemicals. Wear appropriate eye protection, gloves, and protective attire. The use of concentrated acids and bases should be done in the fume hood.

14.4.2 Accessories to Be Used with the GC per Group

One digital flow check meter.

One GC syringe with a beveled end that includes a Chaney adapter. Do not confuse this syringe with the blunt-end syringe used for HPLC.

One GC test mix for each of the studies discussed below.

14.4.2.1 Summary of Turbochrom Methods to Be Used in this Experiment

Order	Turbo Method	Remarks
1	FLOWRATE	Near-ambient column temperature
		Measure flow rate, split ratio
2	DETSENS	Neat acetone—FID
		Neat methylene chloride—ECD
3	INJECD	Inject increasing amounts of 10 ppm 1,2,4-trichlorobenzene
	INJFID	Inject increasing amounts of hexadecane at 225° C
4	PLATES	Temperature program: 265° C (0.1) to 285° C (10.0) at 6° C/min; multi-component organochlorine test mixture
5	TMAX	Isothermal at 285° C
	TMIN	Isothermal at 200° C

14.4.3 Procedure

Refer to "Summary of Turbochrom Methods" for a definition of each of the Turbochrom methods created in support of this experiment. These previously created methods are illustrative of how chromatography-based software can be used to teach fundamental principles of GC.

14.4.3.1 Measurement and Adjustment of Carrier Gas Flow Rate and Split Ratio

As you approach the gas chromatograph, you will find it in an operational mode, with carrier gas flowing through both capillary columns. If not already set up, retrieve the Turbochrom file titled "FLOWRATE" and download this method. Your first task will be to measure the flow rate of the carrier gas through both capillary columns with the makeup gas off. After turning the makeup gas on, measure the split ratio through the capillary injector using the digital flow check meter. *Record flow rate data in your lab notebook.*

Once the optimum carrier flow rates have been established, the dual detector method titled "DETSENS" can be retrieved from the Turbochrom software, then transferred to the instrument via the interface (a process known as download) and the comparison of detector sensitivity can be undertaken. *Ignite the FID (refer to the operator's manual for the Autosystem GC from PerkinElmer for the specific procedure).*

14.4.3.2 Comparison of the FID vs. the ECD Sensitivity

Allow time for the GC to equilibrate at the column temperature set in the method. Using the manual injection syringe (GC syringes are manufactured by the Hamilton Co. as well as by others), inject equal microliter (μL) aliquots of acetone into both injectors. Observe the appearance of a retained chromatographic peak found in both chromatograms. You cannot assume that t_R will be identical on both columns! Compare the peak heights from both chromatograms. Inject equal μL aliquots into both injectors, as earlier, of the specific chlorinated hydrocarbon that is available. *Record your observations* and compare the peak heights as done previously. Each member of the group should have an opportunity to make these sample injections so as to gain some experience with manual syringe injection of organic solvents.

14.4.3.3 Injection Volume vs. GC Peak Shape

Retrieve the Turbo file titled "INJECD" and download. Inject a series of increasing μL aliquots of a reference solution labeled "10 ppm 1,2,4-trichlorobenzene" into the front capillary injection port. *Observe and record the changes in chromatographic peak shape as the amount of analyte is increased.* Retrieve the Turbo file titled "INJFID" and download. Repeat the series of injections as before using the reference solution labeled "hexadecane" and make these injections into the rear injector. *Observe and record the changes in chromatographic peak shape as the amount of analyte is increased.*

14.4.3.4 Flow Rate vs. Capillary Column Efficiency

A column's efficiency is determined in a quantitative manner from the chromatogram by measuring the number of theoretical plates, *N*. The effect of carrier flow rate on capillary column efficiency is significant in GC and will be examined under isothermal conditions (i.e. at a fixed and unchanging column temperature). Retrieve the Turbo file "PLATES" and program this method for a high flow rate by increasing the head pressure. Save this change in the method and download the method. Turn off the makeup gas and adjust the actual pressure so as to nearly match that which is set in the method and *measure* the flow rate. Turn the makeup back on. Inject 1 μL of the test mix and *observe* the chromatogram. Retrieve the method a second time and reprogram the head pressure to a much lower value. Turn off the makeup, decrease the carrier head pressure, *measure* the new flow rate, turn the makeup gas back on, and then make a second injection using the same volume.

For the carrier gas flow rate that exhibited the highest efficiency, *calculate* the number of theoretical plates using equations from your text. In addition, for the optimum carrier flow rate, choose any pair of peaks and calculate the resolution for that pair.

14.4.3.5 Column Temperature vs. Capacity Factor

Retrieve the Turbo file titled "TMIN" and download the method. Inject approximately 1 μL of the multi-component organochlorine test mix at this column temperature of 200° C. *Observe the degree of separation among organochlorine analytes and record your qualitative comments.*

Retrieve the Turbo file titled "TMAX" and download the method. Inject the same volume of the multi-component organochlorine test mix and *observe* the chromatogram when the column temperature has been increased to 285° C.

14.5 FOR THE LAB NOTEBOOK

Write a brief discussion on how your experimental observations connect to the theoretical relationships for GC introduced in various textbooks and journal articles.

Address the following:

1. Explain why different GC detectors have different instrument detection limits.
2. If you operated a GC at significantly reduced carrier gas flow rates, predict what you would observe in a gas chromatogram for the injection of organic compounds. What would be the principal cause for these observations?
3. Explain why a symmetrical peak shape is important in gas chromatography.
4. What happens to K_D for a given organic compound when column temperature is varied?
5. How efficient is your GC column? That is, what is the number of theoretical plates? How many plates per meter do you have?
6. How is the phase ratio, β, determined for capillary GC columns?

14.6 SUGGESTED READINGS

To develop this experiment, the author consulted the following resources:

Sawyer D., W. Heineman, J. Beebe. *Chemistry Experiments for Instrumental Methods.* New York: John Wiley & Sons, 1984, pp. 321–343.

A thorough grounding in the principles and practice of the GC and GC-MS determinative techniques can be found among others in the resources shown below:

Budde W. *Mass Spectrometry: Strategies for Environmental and Related Applications.* Washington, DC: American Chemical Society, Oxford University Press, 2001.

Grob R., E. Barry, Eds. *Modern Practice of Gas Chromatography.* 4th edn. Hoboken, NJ: Wiley-Interscience, 2004.

Jennings W. *Analytical Gas Chromatography.* San Diego, CA: Academic Press, 1987.

McNair H., J. Miller. *Basic Gas Chromatography.* New York: Wiley Interscience, 1998.

Perry J. *Introduction to Analytical Gas Chromatography.* New York: Marcel Dekker, 1981.

15 Screening for the Presence of BTEX in Wastewater Using Liquid–Liquid Extraction (LLE) and Gas Chromatography

Screening for THMs in Chlorine-Disinfected Drinking Water Using Static Headspace (HS) Gas Chromatography

15.1 BACKGROUND AND SUMMARY OF METHOD

Two analytical *screening* approaches are introduced in this experiment. A suitable extracting solvent is experimentally selected and used to *extract suspected gasoline-tainted water samples* using a mini-LLE sample prep technique in order to detect the presence of BTEX. Alternatively, a simulated chlorinated disinfected drinking water sample is screened for *organochlorine-containing* VOCs using *static headspace gas chromatography* with electron-capture detection (HS-GC-ECD) to detect the presence of trihalomethanes (THMs). Molecular structures together with names for the six BTEX VOCs are shown here:

DOI: 10.1201/9781003260707-15

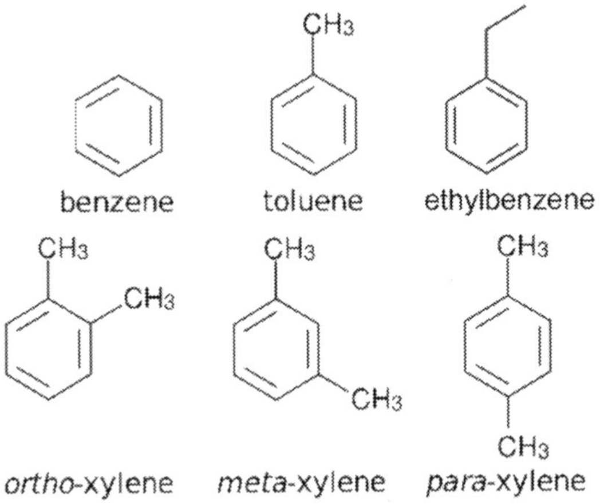

benzene toluene ethylbenzene

ortho-xylene meta-xylene para-xylene

THMs include chloroform, bromodichloromethane, dibromochloromethane, and bromoform. These toxic VOC analytes have been found in drinking water that has been disinfected using chlorine. Molecular structures for the four THMs are shown here:

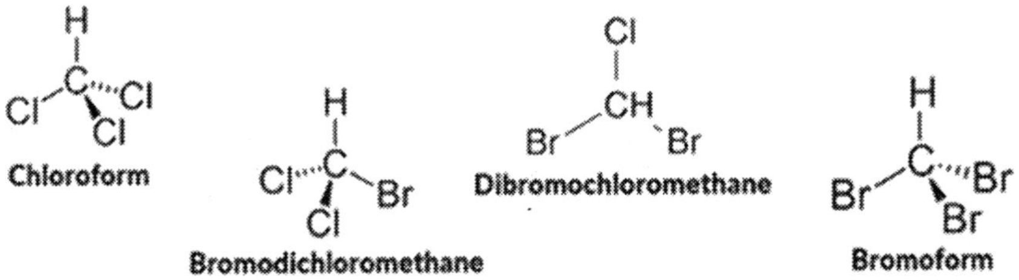

Chloroform Bromodichloromethane Dibromochloromethane Bromoform

We will take a more simplified approach to trace VOCs analysis, which utilizes our limited sample preparation and instrumentation capabilities in the instructional laboratory. If a suitable extraction solvent can be found, i.e. one that does not interfere with the VOCs to be identified and quantitated by gas chromatography, then the analytes of interest can be isolated and concentrated from the environmental sample matrix via a mini-LLE technique. A 40 mL sample of wastewater that might contain BTEX is extracted with 2 mL of a suitable organic solvent. The organic solvent, being less dense than water, conveniently occupies the neck of a 40 mL vial. A 2 μL aliquot of the extract is taken by a liquid-handling syringe and injected into a C-GC-FID to screen for the presence of BTEX compounds. The C-GC-FID must be previously optimized to separate most BTEX compounds. In a separate experiment, 40 mL of chlorine-disinfected drinking water is placed in a sealed HS vial, heated, and 0.5 cc of the headspace is sampled using a HS sampling syringe and injected directly into a C-GC-ECD. The C-GC-ECD must be previously optimized to separate the four THMs.

Typical levels of BTEX contamination for wastewater are in the low parts per million (ppm) concentration range. Typical levels of THM contamination for chlorine-disinfected drinking water are typically between 10 and 100 ppb for each THM. A severe limitation to LLE techniques is the possible formation of emulsions when applied to wastewaters that could have an appreciable surfactant concentration level. HS-C-GC-ECD is a very selective approach for screening chlorine-disinfected drinking water samples for THMs.

15.2 OF WHAT VALUE IS THIS EXPERIMENT?

This exercise affords the student an opportunity to further utilize gas chromatography, this time as a screening tool. Two different sample preparation approaches to screening are introduced for two somewhat different sample matrices. If a method involves phase distribution equilibria either for screening or for quantification, some analyte will always be lost between phases. Volatility losses can be considerable when VOCs are dissolved in water, while these losses are not so critical for SVOCs dissolved in water.

A previously created method will be retrieved from the Turbochrom® (PE Nelson) or other chromatography processing software available in the lab. It is possible for your instructor to turn this qualitative screening experiment into a quantitative determination one. If so, external or internal standards must be prepared and run in order to create the necessary calibration plots.

15.3 EXPERIMENTAL

15.3.1 Preparation of Chemical Reagents

Note: all reagents used in this analytical method contain hazardous chemicals. Wear appropriate eye protection, gloves, and protective attire. The use of concentrated acids and bases should be done in the fume hood.

15.3.2 Chemicals/Reagents Needed per Group

One neat benzene.
One neat toluene.
One neat ethylbenzene.
One neat xylene.
One neat hexane to evaluate as a suitable screening extractant.
One neat hexadecane to evaluate as a suitable screening extractant.
One neat dichloromethane to evaluate as a suitable screening extractant.
One approximately 5,000 ppm stock BTEX standard (refer to actual label for exact values).
One 40 mL of a wastewater sample for screening for BTEXs.

One 40 mL of a chlorine-disinfected drinking water sample for screening for THMs.
One 500 ppm reference stock standard containing THMs in MeOH.

15.3.3 Items/Accessories Needed per Student or per Group

One 42 mL glass vial with screw caps and PTFE/silicone septa.
One 22 mL glass headspace vial with PTFE/silicone septa and crimp-top caps.
One crimping tool for headspace vials.
One liquid-handling syringe whose capacity is 10 μL with a Chaney adapter (Hamilton or another manufacturer) for injection of liquid extracts.
One 0.5 or 1.0 cc capacity gas-tight syringe for headspace sampling and direct injection (Precision Sampling, SGE, and Hamilton, among others, manufacture such syringes).
One heating block assembly that accepts a 22 mL HS vial and allows for measurement of the block temperature (VWR or another supply house).

15.3.4 Preliminary Planning

At the onset of the laboratory period, assemble as a group and decide who is going to do what. Assign specific tasks to each member of the group. Once all results are obtained, the group should reassemble and share all analytical data.

15.3.5 Procedure for BTEX Instrumental Analysis Using Mini-LLE Techniques

15.3.5.1 Selecting the Most Suitable Extraction Solvent

Place one small drop of each of the neat BTEX liquids into approximately 10 mL of hexane. Inject 1 μL into the GC-FID and interpret the resulting chromatogram. Repeat for dichloromethane and then for hexadecane. Methods must be previously created on Turbochrom or equivalent software. Recall, the most suitable solvent is the one that does not interfere with the GC elution of BTEXs. From these observations, select the most appropriate extraction solvent, then proceed to prepare calibration standards.

15.3.5.2 Preparation of the Primary Dilution Standard and Working Calibration Standards

1. Using a clean and dry glass pipette (volumetric), transfer 1.0 mL of the 5,000 ppm BTEX to a 10 mL volumetric flask that has been previously half-filled with the most suitable solvent that you chose earlier. Adjust to the calibration mark with this solvent and label it as "500 ppm BTEX," for example. This is what EPA methods call a primary dilution standard since it is the first dilution that the analyst prepares from a given source. In this case, a 1:10 dilution has been made.

2. Prepare a series of working calibration standards according to the following table:

Standard#	500 ppm BTEX (mL)	Final Volume (mL)	Concentration of BTEX (ppm)
1	1	10	50
2	2	10	100
3	4	10	200
4	8	10	400
5	—	—	500
ICV	5	10	250

For example, to prepare standard 3, transfer 4 mL of 500 ppm BTEX (MeOH) to a 10 mL volumetric flask half-filled with MeOH. Add sufficient MeOH to adjust the meniscus to the mark of the volumetric flask. This yields a calibration standard whose concentration is 200 ppm BTEX dissolved in MeOH.

3. Retrieve the method BTEX from the Turbochrom software, open a new sequence file, and name the raw data file in a manner similar to the following example: "G116" (Group 1, 16th of the month). Save the sequence file and name it in a manner similar to the following example: "G10316" (Group 1, March 16).
4. Inject 1 μL aliquots of all calibration standards and inject the ICV in triplicate. Update the calibration method within the Turbochrom software. *Ask your lab instructor for help in updating the calibration within the method.* Observe the calibration curve and note the value of the square of the correlation coefficient. Discuss with your instructor whether this calibration is acceptable.
5. After the instrument has been properly calibrated and the ICVs quantitatively determined, proceed to inject the unknown sample extracts from the mini-LLE (refer to "Procedure to Conduct a Screen ..." below). Obtain the interpolated values from the external standard mode of instrument calibration.

15.3.6 Procedure for THM Instrumental Analysis Using HS Techniques

Using the 500 ppm THM stock reference solution, prepare a series of calibration standards in which the THMs are present in 10 mL of DDI, which is contained in a 22 mL HS vial with PTFE/silicone septa and aluminum crimp-top caps. *Refer to the BTEX calibration for guidance* as you prepare a series of working calibration standards for HS-GC analysis. Ask your instructor to review your calibration table for correctness. Following the development of a calibration curve, inject the ICV (only one injection per sample is acceptable in HS-GC), then inject the headspace above the aqueous samples. Following the development of a calibration curve, inject the ICV and the chlorine-disinfected drinking water samples.

15.3.7 Procedure to Conduct a Screen for BTEXs via Mini-LLE and Subsequent Injection into a GC-FID

Once the most appropriate extraction solvent has been selected, the wastewater sample that contains dissolved BTEX can be extracted. To a clean 42 mL glass vial with a PTFE/silicone septum and screw cap, add 40 mL of aqueous sample. Pipette 2.0 mL of extraction solvent, and place the septum and cap in place. Shake for 1 min and let stand for at least 5 min until both phases clearly separate. Using a glass transfer pipette, remove approximately 75% of the extract and place it in a small test tube or vial. Inject 1 μL of extract into the GC-FID. *Discard the contents of the 42 mL glass vial into the waste receptacles that are located in the laboratory.*

15.3.8 Procedure to Conduct Manual Headspace Sampling and Direct Injection into a GC-ECD

If a heater block is available, place the sealed and capped 22 mL headspace vial into the block and allow time for the vial to equilibrate before sampling. A water bath, i.e.,a large beaker that is half-filled with water, could serve as a constant-temperature environment for headspace sampling. Insert the gas-tight syringe with the valve in the "on" position (applicable to syringes made by VICI Precision Sampling and others) by penetrating the septum seal, and withdraw a 0.5 cc aliquot of headspace. Be careful not to withdraw any liquid. Immediately close the on–off valve to the syringe while positioned within the headspace. Position the syringe into the injection port, open the syringe valve, and transfer the 0.5 cc aliquot into the GC.

15.4 FOR THE REPORT

Since this experiment involves screening only, quantification of the wastewater and chlorine-disinfected samples is unnecessary unless your instructor asks you to quantitate. If you find BTEX or THMs from the screens, discuss how you might conduct a quantitative analysis of these samples. If these samples identified additional compounds that were not BTEX or THM compounds, suggest ways that the identity of these unknown compounds could be revealed. Explain the basis on which you chose the screening extractant. Refer to literary resources that discuss how to screen for VOCs for some help here.

15.5 SUGGESTED READINGS

To develop this experiment, the author consulted the following resources:

Grob R., E. Barry, Eds. *Modern Practice of Gas Chromatography.* 4th edn. Hoboken, NJ: Wiley-Interscience, 2004, pp. 563–573.

Kolb B., L. Ettre. *Static Headspace-Gas Chromatography: Theory and Practice.* New York: Wiley VCH, 1997.

Sawyer D., W. Heineman, J. Beebe. *Chemistry Experiments for Instrumental Methods.* New York: Wiley, 1984, pp. 321–343.

16 Determination of Priority Pollutant Volatile Organic Compounds (VOCs) in Gasoline-Contaminated Groundwater Using Static Headspace (HS) and Solid-Phase Microextraction Headspace (SPME-HS) and Gas Chromatography

16.1 BACKGROUND AND SUMMARY METHOD

Benzene, toluene, ethylbenzene, *para-*, *meta-*, and *ortho*-xylenes, collectively referred to as BTEX, constitute some of the most environmentally detrimental organic compounds that have made their way into groundwater, primarily due to gasoline spills and underground rusted storage tank leakage over time. In preparing for this experiment, this author in the 1990s (for educational purposes) contaminated (in the teaching laboratory) a small sample volume of groundwater with gasoline. The author observed not only the six BTEX compounds, however; in addition, he observed a large and early eluting peak that matches the retention time of methyl-*tert*-butyl ether (MTBE). MTBE was used as a gasoline additive designed to boost octane rating and was thought at the time to be a suitable substitute for tetraethyl lead. However, MTBE quickly polluted groundwater as well! BTEX compounds comprise about 20 to 30% of gasoline and have appreciable solubility in water in contrast to aliphatic hydrocarbons such as n-heptane, iso-octane, and n-dodecane. The Energy Policy Act of 2005 prompted gasoline refiners to transition away from

DOI: 10.1201/9781003260707-16

MTBE to the use of ethanol as a gasoline additive. Molecular structures for MTBE, tetraethyl lead, and ethanol are shown here:

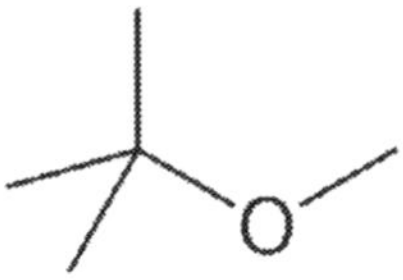

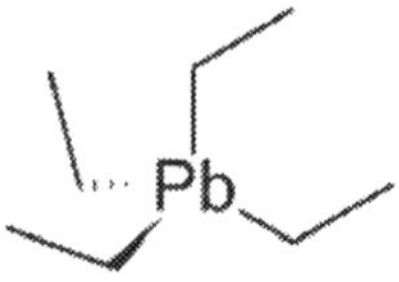

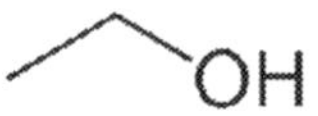

MTBE Tetraethyl lead Ethanol

EPA Methods 601 (purgeable halocarbons) and 602 (purgeable aromatics) comprise the real workhorse approaches to trace VOCs analyses in wastewaters. These methods use dynamic headspace sampling coupled to GC with electrolytic conductivity (EPA Method 601) and photoionization (EPA Method 602) detection. EPA Method 502.2 is a high-resolution capillary column method with both detectors cited above connected in series and provides monitoring capabilities for over 60 VOCs that could be found in municipal drinking water supplies. An alternative to dynamic headspace (commonly referred to as purge-and-trap) is static headspace capillary gas chromatographic (HS-C-GC) techniques.

Static HS techniques take advantage of the volatility exhibited by VOCs whereby the air remaining in a sealed vial above a liquid (defined as the headspace) is sampled with a gas-tight syringe and directly injected into the GC-FID for carbon-containing VOCs. This technique is called *manual HS-GC*, as distinguished from automated HS-GC techniques. A complement to static HS is *SPME-HS*. A fiber coated with a polymer such as poly-dimethyl siloxane is inserted through the septum and into the HS. VOCs partition from the HS to the polymer film. The SPME syringe-fiber assembly is removed from the vial and inserted directly into the injection port of a gas chromatograph. VOCs are thermally desorbed off of the fiber by the hot GC inlet and onto the head of a wall-coated open tubular (WCOT) column where the VOCs are gas chromatographed.

In commercial laboratories, HS-GC and SPME-HS-GC sample prep techniques have been automated! The CombiPal® (CTC Analytics) offers a HS syringe mounted on a robotic head that moves horizontally (known as a rail). This robotic autosampler can accommodate either a *gas-tight syringe* to conduct automated static HS-GC sampling or an *SPME syringe* holder to conduct automated SPME-HS-GC. The robotic autosampler called a *Multipurpose Sampler* (MPS) is complemented and controlled through hardware and software provided by Gerstel GmbH and Co. KG and others. A second rail on the MPS provides robotic automated reagent delivery to the HS vial.

16.2 OF WHAT VALUE IS THIS EXPERIMENT?

Students will have an opportunity in this experiment to quantitatively determine the concentration level of various BTEX compounds from gasoline-contaminated groundwater using both static HS and SPME-HS techniques. Both sampling/sample prep techniques will be performed *manually*. This experiment affords students an opportunity to operate a conventional gas chromatograph. This GC is interfaced with a personal computer that utilizes Turbochrom® (PerkinElmer) software or the equivalent for data acquisition. Hence, a student must become familiar with the sampling/sample prep technique, the GC, and the software used at the same time. This trio of techniques comprises a necessary learning experience for the student who will eventually work in fields related to trace environmental quantitative analysis.

The method titled "BTEX.mth" will be retrieved from Turbochrom or other chromatography processing software available in the lab. External standard calibration curves will be generated using both HS-C-GC-FID and SPME-HS-C-GC-FID methods. An aqueous environmental sample that has been contaminated with gasoline will be available and analyzed for traces of BTEX. Since two different analytical methods are applied to the same standards and samples, students will have the opportunity to apply t statistics to compare the analytical results from both methods.

16.3 USE OF *t* STATISTICS

Comparison of two dependent averages is a statistical procedure that helps to determine whether two different analytical methods give the same average result for a given sample. If one analyzes each of a series of samples, which could include calibration standards, ICVs, blanks, and unknowns, by the two methods, *a pair of results* for *each sample* will be obtained. The difference between these two results for each pair will reflect only the difference in the methods. The following equations are used for the t test on paired data:

$$t_{\text{calc}} = \frac{\bar{d}}{S_d}\sqrt{n}$$

$$S_d = \sqrt{\frac{\sum d^2 - \frac{\left(\sum d\right)^2}{n}}{n-1}}$$

where

d = difference in each pair of values

$\bar{d}$ = average *absolute* difference in the pairs of *values*

n = number of pairs of values

df = degrees of freedom associated with a given value for t

s_d = standard deviation of the differences between the pairs of observations.

A comparison of the calculated value, t_{calc}, with that from a *table of student's t values* is then made. If $t_{\text{calc}} > t$ (from a student's t table at the desired level of significance), then both methods do not give the same result. If $t_{\text{calc}} < t$ (from a student's t table at the desired level of significance), then it is statistically valid to assume that both methods are equivalent.

16.4 EXPERIMENTAL

16.4.1 Preparation of Chemical Reagents

Note: all reagents used in this analytical method contain hazardous chemicals. Wear appropriate eye protection, gloves, and protective attire. The use of concentrated acids and bases should be done in the fume hood.

16.4.2 Chemicals/Reagents Needed per Group

One neat benzene.
One neat toluene.
One neat ethylbenzene.
One neat xylene: *ortho*, *meta*, and *para*.
One 2,000 ppm BTEX, certified reference standard, dissolved in MeOH.
One 40 mL of gasoline-contaminated groundwater for BTEX determination.
One 40 mL of an unknown sample prepared by the staff for BTEX determination.

16.4.3 Items/Accessories Needed per Student or per Group

Ten 22 mL glass headspace vials with PTFE/silicone septa and aluminum crimp-top caps.
Seven 10 mL glass volumetric flasks to be used to prepare the calibration and ICV reference standards (refer to the calibration table below).
One crimping tool for headspace vials.
One 0.5 or 1.0 cc capacity gas-tight syringe for headspace sampling and direct injection (manufacturers of syringes include: VICI Precision Sampling, SGE, Hamilton, and others).
One heating block assembly that accepts a 22 mL HS vial and allows for measurement of the block temperature.
One SPME fiber holder for manual sampling (Millipore Sigma, formerly Supelco).
One manual SPME sampling stand (Millipore Sigma, formerly Supelco) or equivalent, including mini stir bars. This apparatus is optional; the heating block assembly can be used to conduct SPME-HS.
One 100 μm (film thickness) of poly-dimethyl siloxane (PDMS) fiber for use with the SPME holder. Instructions for installing the PDMS fiber into the SPME fiber holder are available from the manufacturer (Millipore Sigma, formerly Supelco).

16.4.4 Preliminary Planning

At the onset of the laboratory period, assemble as a group and decide who is going to do what. Assign specific tasks to each member of the group. Once all results are obtained, the group should reassemble and share all analytical data.

16.4.5 Procedure for BTEX Instrumental Analysis HS Techniques

Using the 2,000 ppm BTEX stock reference solution dissolved in MeOH, prepare a series of calibration standards in which the BTEX is present in a final volume V_T = 10 mL of DDI. Refer to the table below. Use of a 10 mL volumetric flask to initially prepare reference standards prior to transfer to the 22 mL headspace vial is available in this experiment but optional. Refer to the procedure stated immediately below the calibration table. Transfer each reference standard and each ICV to a 22 mL HS vial and immediately close with a PTFE/silicone septa and aluminum crimp-top cap. Refer to the calibration table below for reference as you prepare a series of working calibration standards for HS-GC analysis. Following the development of a calibration curve, inject the ICV (only one injection per vial is acceptable in HS-GC) and then one or more of the gasoline-contaminated aqueous samples.

- Prepare a series of working calibration standards and ICVs according to the following table:

Standard No.	2,000 ppm BTEX (MeOH) (μL)	Vt (mL)	Concentration of BTEX (ppm)
Blank	0	10	0
1	20	10	4
2	40	10	8
3	80	10	16
ICV 1	30	10	6
ICV 2	30	10	6
ICV 3	30	10	6

- Place the indicated aliquot of 2,000 ppm BTEX (MeOH) into a 22 mL HS vial containing 10 mL of DDI and seal promptly using the crimping tool. Place the vial into the heated block, whose temperature should be elevated above ambient. Maintain this temperature throughout the experiment.
- Retrieve the method BTEX from Turbochrom or equivalent software. Open a new sequence file and name the raw data file according to the following example: "G116" (group 1, 16th of the month). Save the sequence file and name it in a manner similar to that in the following example: "G10316" (group 1, March 16). Download the method (BTEX.mth) and the sequence file (e.g. G10316.seq).
- Make manual injections of approximately 0.25 cc of headspace using a gas-tight syringe (refer to the technique section below). After the three calibration standards have been run, update the calibration method within Turbochrom or equivalent software. *Ask your lab instructor for help in updating the calibration within the method.* Observe the calibration curve and note the value of the square of the correlation coefficient. Discuss with your instructor whether this calibration is acceptable.

- After the instrument has been properly calibrated, proceed to inject the headspace for the three ICVs, a method blank, and unknown samples, as assigned. Your instructor may give you a sample whose concentration is unknown. Record the code on the vial label.
- Obtain the interpolated values from the least squares regression for your three ICVs, method blank, and any and all samples. *Obtain assistance from staff in getting a hard copy of your results.*

16.4.6 Technique to Conduct a Manual Headspace Sampling and Direct Injection Using a Gas-Tight Sampling Syringe

If a heater block is available, place the sealed and capped 22 mL headspace vial into the block and allow time for the vial to equilibrate before sampling. A water bath, i.e. a large beaker that is half-filled with water, could serve as a constant-temperature environment for headspace sampling. Insert the gas-tight syringe with the valve in the "on" position (if a Precision Sampling–type syringe is used) by penetrating the septum seal, and withdraw a 0.25 cc aliquot of headspace. Be careful not to withdraw any liquid. Immediately close the on–off valve to the syringe while positioned within the headspace. Position the syringe into the injection port, open the syringe valve, and transfer the 0.5 cc aliquot into the GC.

16.4.7 Technique to Conduct an SPME Headspace Sampling and Injection/Thermal Desorption Using an SPME Syringe/Fiber Assembly

Install the 100 µm PDMS Fiber into the SPME holder if this has not already been done by your instructor. Follow directions for installing the fiber. Clean the fiber by inserting the SPME holder into an unused GC injection port whose temperature is ~250° C for about ½ hour. To a sealed HS vial that contains either spiked water or an aqueous unknown sample, add a stir bar and begin magnetic stirring. Insert the retracted fiber holder through the septum. Expose the fiber by depressing the plunger and lock it in the bottom position by turning it clockwise. The PDMS fused silica fiber that is attached to a stainless-steel rod is now exposed to the HS. The fiber should remain above the height of the liquid level. Allow the extraction to take place for ~2 min. Re-track the fiber back into the needle and pull the device out of the vial. Insert the needle of the SPME device into the injection port of the GC. This must be done carefully since SPME needles tend to be of a thinner gauge. Start the analysis by depressing the plunger and locking it in position. After 30 sec, withdraw the fiber back into the needle and pull the needle out of the injector. When the separation is completed, repeat the analysis to determine fiber carryover. Repeat this technique for every calibration standard, ICV, blank, and sample in the same general manner introduced above for the static HS technique.

16.5 FOR THE REPORT

For each sample prep technique, include:

1. A three-point external calibration plot for each chromatographically resolved BTEX analyte with a corresponding *correlation coefficient*. Note that Turbochrom finds the square of the correlation coefficient, known as a *coefficient of determination*.
2. A table that includes results for all calibration standards, ICVs, and samples from an interpolation of the regressed calibration curve. This table is to be used for the statistical comparison between both methods.
3. *The coefficient of variation for the ICVs.*
4. The *relative error* for the ICVs.
5. A representative gas chromatogram for the separation.

How do both techniques compare? Apply the *comparison of two dependent averages* to all data. Write several paragraphs using your findings to address this question. Identify those sources of error that might compromise accuracy and precision for both techniques. How might the calibration procedure be modified if an internal standard mode of instrument calibration were used? If an isotope dilution approach were used?

16.6 SUGGESTED READINGS

To develop this experiment, the author consulted the following resources:

Grandy J. et al. Frontiers of sampling: Design of high surface area thin-film samplers for on-site environmental analysis. *LC-GC North America* 37(9): 690–697, 2019. This paper, a recent contribution from the Pawliszyn research group at the University of Waterloo, introduces various designs developed for SPME sampling of a wide variety of environmental sample matrices. Archives of back issues of LC-GC are available online.

Grob R., E. Barry, Eds. *Modern Practice of Gas Chromatography*. 4th edn. Hoboken, NJ: Wiley-Interscience, 2004, pp. 563–573.

Kolb B., L. Ettre. *Static Headspace-Gas Chromatography: Theory and Practice*. New York: Wiley VCH, 1997.

Pawliszyn J. *Solid Phase Microextraction: Theory and Practice*. New York: Wiley VCH, 1997, pp. 193–200. This was the first book published on SPME theory and application written by the inventor of this sample prep technique.

Sawyer D., W. Heineman, J. Beebe. *Chemistry Experiments for Instrumental Methods*. New York: Wiley, 1984, pp. 321–343.

17 Determination of the Herbicide Residue Trifluralin in Chemically Treated Lawn Soil by Gas Chromatography Using Reversed-Phase Solid-Phase Extraction (RP-SPE) Sample Prep Techniques

17.1 BACKGROUND AND SUMMARY OF METHOD

The persistence of trace residue levels of pesticides and herbicides in the environment has been a cause for continued concern since the early 1960s, when it became apparent that these residues were detrimental to wildlife and possibly to human health. The benefits of using DDT gradually gave way to the increasing risk of continued use and eventually led to the banning of its use. Herbicides, however, do not appear to present such a high risk to the environment and continue to find widespread use. The chlorophenoxy acid herbicides are not directly amenable to GC and must first be chemically converted to their more volatile methyl esters prior to analysis using GC. Trifluralin or, according to the International Union of Pure and Applied Chemistry (IUPAC) organic nomenclature, α,α,α-trifluoro-2,6-dinitro-*N*,*N*-dipropyl-*p*-toluidine, is commonly one of the active pre-emergent herbicide ingredients in some lawn treatment formulations. Consider the molecular structures (shown below) of trifluralin and the *internal standard 1,2,4-trichlorobenzene*. The pre-emergent herbicide trifluralin will be extracted from herbicide treated lawn soil and quantitated against 1,2,4-trichlorobenzene in this experiment:

DOI: 10.1201/9781003260707-17

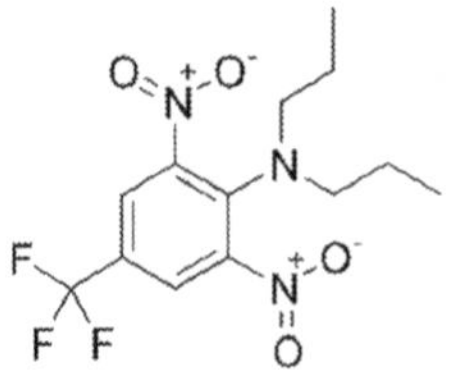

Trifluralin 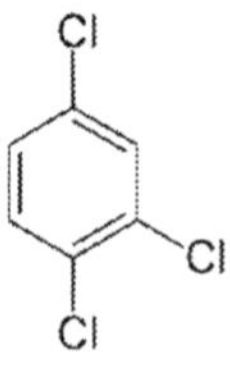 1,2,4-trichlorobenzene

With reference to the molecular structure for trifluralin, the presence of electronegative heteroatoms, such as fluorine combined with two nitro substituents on the benzene ring, would make the organic compound highly sensitive to the electron-capture detector (ECD), provided that the substance is sufficiently vaporizable and therefore amenable to GC. With a boiling point of 139° C, trifluralin is appropriately classified as a semi-volatile, neutral organic compound (SVOC) and it is thus feasible to think that trifluralin could be isolated from a soil matrix by conventional sample preparation techniques such as liquid–liquid extraction (LLE) or perhaps by the more recently developed sample prep technique, reversed-phase solid-phase extraction (RP-SPE).

The assay (on the package) for the commercially available formulation that was dispersed over a lawn whose soil beneath has been sampled is given as follows:

Ingredient	**%**	**Ingredient**	**%**
Total nitrogen	*20*	*Chlorine (not more than)*	*3*
Trifluralin (N,N-dipropyl)	*0.82*	*Ammoniacal nitrogen*	*1.17*
Urea nitrogen	*18.83*	*Trifluralin (N-butyl, N-ethyl)*	*0.43*
Sulfur	*1.2*	*Soluble potash*	*3*
Available phosphate	*3*	*Inert*	*98.5*

17.1.1 Solid-Phase Extraction

Reversed-phase solid-phase extraction (RP-SPE) techniques provide an alternative to LLE whereby a chemically bonded silica gel is packed into 3 mL barrel-type cartridges (cylindrical) or impregnated into disks that are in turn packed in the same 3 mL barrel cartridges and used to isolate and recover SVOCs contaminants from various environmental samples. A chemically neutral (non to moderately polar) organic compound originally dissolved in water at trace concentration levels is thermodynamically unstable. If an aqueous solution containing this compound is allowed to contact a hydrophobic surface, a much stronger van der Waals type of intermolecular interaction *causes the molecules of the analyte to adsorb or stick* to the surface, and thus effectively get removed from the aqueous solution. A relatively small volume of a nonpolar or even semi-polar solvent provides enough hydrophobic interaction to then remove (elute in a chromatographic sense) the analyte molecules. The following sketch is a schematic for the interaction of analyte molecules 2-naphthylamine and hexyl benzene sulfonate (isolates), with a C_8-bonded silica surface:

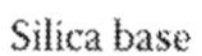

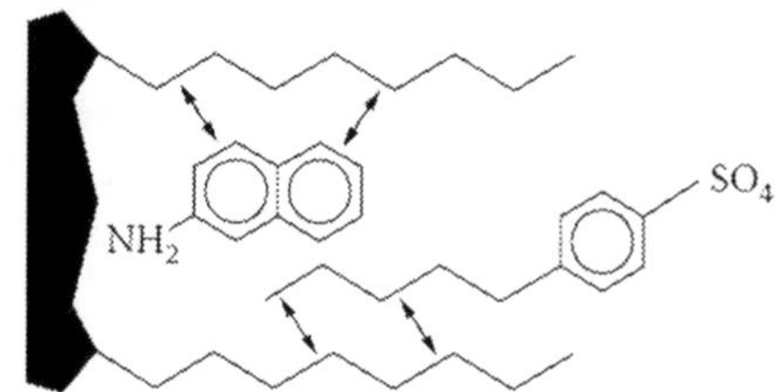

The RP-SPE technique is performed in a stepwise manner as follows:

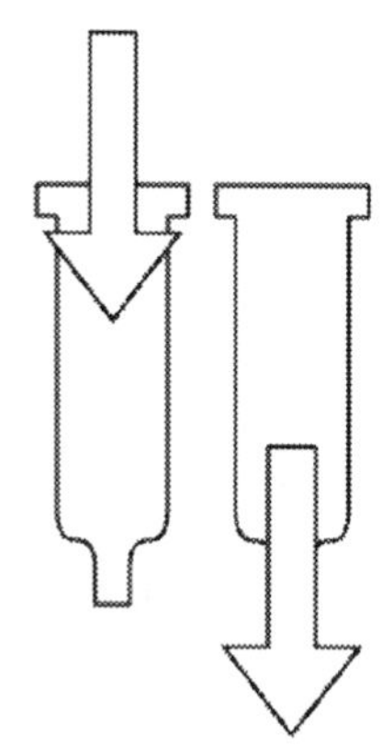

Conditioning

Conditioning the sorbent prior to sample application ensures reproducible retention of the compound of interest (the isolate).

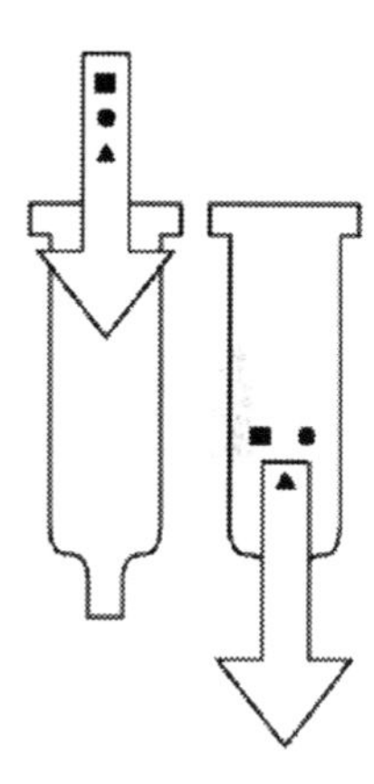

Retention

■ Adsorbed isolate

□ Undesired matrix constituents

Δ Other undesired matrix components

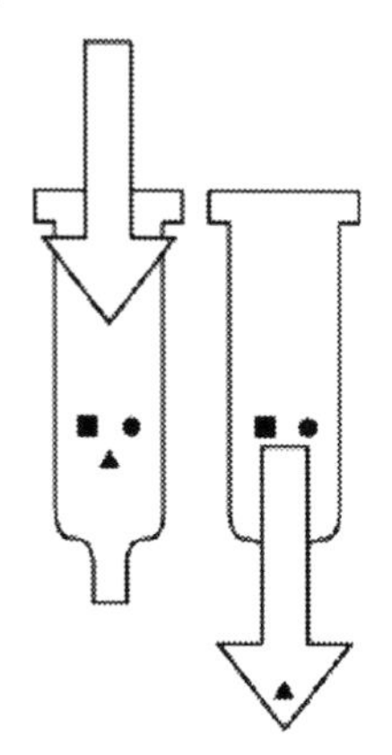

Rinse

▲ Rinse the columns to remove undesired matrix components

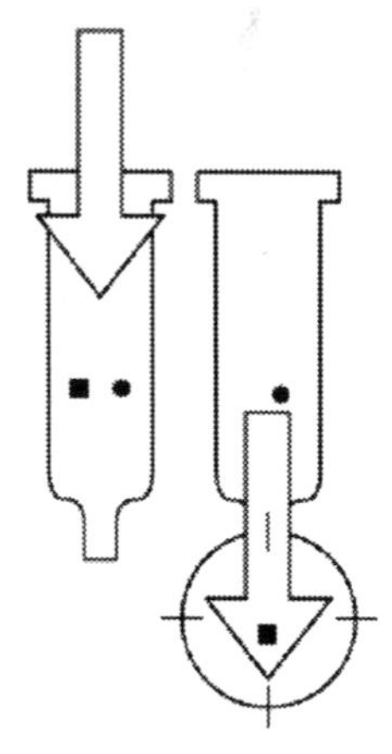

Elution

■ Purified and concentrated isolate

□ Undesired components remain

Trifluralin, which might be present in lawn-treated soil, will be initially extracted into methanol. The methanol extract will be diluted with distilled deionized water (DDI), and the aqueous solution transferred to a 70 mL SPE reservoir on top of a conditioned C_{18}-bonded silica sorbent. The sorbent cartridge will be eluted with high-purity (often called *pesticide residue grade*) iso-octane. The iso-octane eluent is dried by passing it through a second SPE cartridge that contains *anhydrous sodium sulfate* (Na_2SO_4) directly into a 1.0 mL volumetric receiver. An *internal standard* is then added and the eluent brought to a final volume of exactly 1.0 mL. A 1 to 2 μL aliquot of the eluent can then be directly injected in a C-GC-ECD (Autosystem® GC PerkinElmer). The concentration of trifluralin in the eluent can be determined after establishing and verifying the instrument calibration.

17.1.2 Internal Standard Mode of Calibration

In addition to external standard and standard addition, the last principal mode of calibration is the *internal standard*. This mode of calibration should be used when there exists *variability in sample injection volume* or when there is concern about the *lack of instrument stability*, or when there is *unavoidable sample loss*. Instrumental response becomes related then to the ratio of the unknown analyte X to that for the internal standard S, instead of becoming related only to the unknown analyte. If some X is lost, one can assume that some S would be lost as well. This preserves the ratio [X]/[S]. For extraction methods, the internal standard (IS) is added to the final extract *just prior to adjusting the final volume. Selecting a suitable IS is not trivial.* The IS should possess similar physical and chemical properties to the analyte of interest and not interfere with the elution of any of the analytes that need to be identified and quantitated. The IS should be within the same concentration range as for the calibration standards and at a fixed concentration. A calibration curve for the IS mode is shown here:

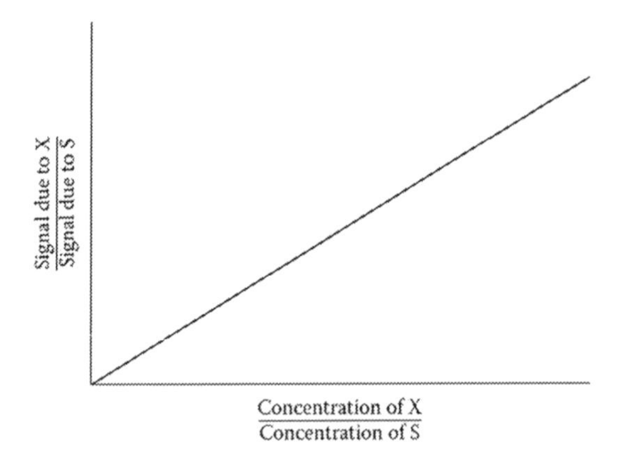

17.2 EXPERIMENT

17.2.1 Preparation of Chemical Reagents

Note: all reagents used in this analytical method contain hazardous chemicals. Wear appropriate eye protection, gloves, and protective attire. The use of concentrated acids and bases should be done in the fume hood.

17.2.2 Chemicals/Reagents/Accessories Needed per Group

One 10 mL of iso-octane, suitable for trace pesticide residue analysis.
One 100 mL of methanol, suitable for trace pesticide residue analysis.
Ten SPE cartridges packed with approximately 200 mg of C_{18}-bonded silica.
Ten empty SPE cartridges loosely packed with approximately 500 mg of anhydrous sodium sulfate.
One SPE vacuum manifold connected to a vacuum pump via a water trap.
Ten 1.0 mL glass volumetric flasks with ground-glass stoppers.
One 10 μL syringe (Hamilton or others) for injection into the GC.
One 10 ppm trifluralin reference stock standard dissolved in *pesticide residue grade* iso-octane.
One 10 ppm 1, 2, 4-trichlorobenzene (IS) dissolved in MeOH or methyl-*tert*-butyl ether (MTBE).

17.2.3 Preparation of the Working Calibration Standards

From the reference stock solution of trifluralin in iso-octane, prepare a series of working calibration standards that cover the range of concentration levels between 10 and 1,000 ppb of trifluralin in high-purity iso-octane. Each working standard should also contain the IS at a concentration level that should fall within the range of concentrations for the calibration standards. *This level should be identical among all standards and sample extracts*. Use the table below to guide you in preparing your calibration standards.

Standard #	10 ppm Trifluralin (μL)	10 ppm IS (μL)	$V(T)$ (mL)	Concentration of Trifluralin (ppb)
0	0	50	1.0	0
1	25	50	1.0	250
2	50	50	1.0	500
3	100	50	1.0	1,000
ICV	40	50	1.0	400

17.2.4 Establishing the Calibration

Retrieve the method from Turbochrom® or other equivalent software titled "Triflu.mth," *create* a sequence file, and *download* the sequence. *Turn off* the nitrogen makeup gas to the ECD and *measure the split ratio. Adjust to a ratio of between 15 and 20 to 1. Turn the makeup on* after you make the split ratio measurements.

Inject 1 μL of each working calibration standard and inject the ICV in triplicate using the 10 μL liquid-handling syringe. *Update* the method titled "Triflu.mth" with the new calibration standards data using the Graphic Editor. The precision and accuracy data for the ICV can be obtained by retrieving the Graphic Editor and bringing up the raw file for each ICV. Print the tabular formatted report for that particular sample.

17.2.5 Isolating Trifluralin from Lawn-Treated Soil Using RP-SPE Techniques

An SPE vacuum manifold, which should be connected to a vacuum pump via a water trap, should be available at the workbench for each of the four workstations. *Condition the C_{18} sorbent by passing 2 mL of MeOH through it.* Attach a 70 mL polypropylene reservoir to the top of the SPE cartridge and fill with DDI to approximately two-thirds full.

Place 0.25 g of lawn-treated soil into each of three 50 mL beakers, add 10 mL of methanol (pesticide residue grade) to each beaker, and use a glass stirring rod or magnetic stirring bar to stir this mixture for 5 min. Let stand for another 5 min; then *decant the supernatant liquid* through a Pasteur pipette, which contains non-silanized glass wool to remove large particulates, into a clean beaker. Transfer the liquid to the 70 mL reservoir. Turn on the vacuum pump and pass the contents of the reservoir through the C_{18} sorbent cartridge. After the contents of the reservoir have passed through the sorbent, rinse the reservoir and cartridge with DDI.

Remove the surface moisture with a tissue or equivalent and attach a second SPE cartridge that contains *anhydrous sodium sulfate* beneath the C_{18} sorbent cartridge that contains the retained trifluralin. Elute the sorbent with two 500 μL aliquots of high-purity iso-octane into a 1.0 mL volumetric flask. Remove the receiving volumetric flask from the apparatus and adjust to the calibration mark on the flask with iso-octane.

Inject 1 μL of the dried iso-octane eluent into the C-GC-ECD. Repeat for the other two samples.

17.3 FOR THE REPORT

Include all calibration plots, correlation coefficients, and precision and accuracy results of the ICV, and report on the concentration of trifluralin in the soil in #mg/kg (ppm).

17.4 SUGGESTED READING

To develop this experiment, the author consulted the following resources:

Hagen D., C. Markell, and G. Schmidt. Membrane approach to solid-phase extraction. *Anal Chim Acta* 236: 157–164, 1990. This paper was the first to report on a new RP-SPE sorbent, the Empore® Disk (3M Corp).

Lee H., A. Chau, F. Kawahara. Organochlorine pesticides. In Chau A., B. Afghan Eds. *Analysis of Pesticides in Water*. Vol II. Boca Raton, FL. CRC Press, 1982. Chapter 1. This work is part of a three-volume series and is a good introduction to pesticide residue analysis even though it is outdated in the use of packed, instead of capillary GC columns.

Loconto P.R. Solid-phase extraction in trace environmental analysis, Part I. *LC-GC* 9: 460–465, 1991; Loconto P.R. Solid-phase extraction in trace environmental analysis, Part II. *LC-GC* 9: 752–760, 1991. The author's pioneering work demonstrated that the multi-component EPA environmental methods for SVOCs were adaptable to a multi-modal RP-SPE approach. Archives of back issues of LC-GC are available online.

Methods for the Determination of Organic Compounds in Drinking Water. EPA-600/4-88/039. Cincinnati, OH: Environmental Monitoring Systems and Support Laboratory, December 1988.

Perry J. *Introduction to Analytical Gas Chromatography*. New York: Marcel Dekker, 1981. This is one of the better discussions of the principles behind the operation of the ECD up through technological development at that time.

Sawyer D., W. Heineman, J. Beebe. *Chemistry Experiments for Instrumental Methods*. New York: Wiley, 1984, pp. 321–343.

18 Determination of Priority Pollutant Semivolatile Organochlorine Pesticides

A Comparison of Mini-Liquid–Liquid and Reversed-Phase Solid-Phase Extraction Techniques

18.1 BACKGROUND AND SUMMARY OF METHOD

Organochlorine pesticides (OCs) were used widely in agriculture during the first half of the 20th century in the US and were subsequently banned from use during the 1970s. Unfortunately, some of the OCs like DDT are still in widespread use around the world. Their persistence in the environment was not apparent until Lovelock introduced the electron-capture detector (ECD) in 1960. When combined with high-resolution capillary gas chromatography and appropriate sample preparation methods, the ECD provides the analytical chemist with the most sensitive means by which to identify and quantitate OCs in environmental aqueous and soil/sediment samples. As analytical chemists were seeking to identify and quantitate OCs during the early 1970s, it became apparent that many additional chromatographically resolved peaks were appearing. What were considered as unknown interfering peaks in the chromatogram were then subsequently found to be polychlorinated biphenyls (PCBs)!

The OCs and PCBs were first determined in wastewaters using EPA Method 608. This method originally required packed columns, and because of this, it necessitated extensive sample preparation and cleanup techniques, which included liquid–liquid extraction and low-pressure column liquid chromatography. Capillary GC-ECD, when combined with more contemporary methods of sample

DOI: 10.1201/9781003260707-18

preparation, provides for rapid and cost-effective trace environmental analysis. Over the past 30 years, there have been dramatic improvements in sample preparation techniques as they relate to semivolatile and nonvolatile trace organics quantitative analyses.

In addition to external standard and standard addition, the last principal mode of calibration is called *internal standard*. This mode of calibration should be used when there exists variability in sample injection volume, when there is concern about the lack of instrument stability, and when there is unavoidable sample loss. The instrumental response then becomes related to the ratio of the unknown analyte X to that for the internal standard S, instead of related only to the unknown analyte. If some X is lost, one can assume that some S would be lost as well. This preserves the ratio [X]/[S]. For extraction methods, the internal standard (IS) is added to the final extract just prior to adjusting the final volume. *Selecting a suitable IS is not trivial.* It should possess similar physical and chemical properties to the analyte of interest and not interfere with the elution of any of the analytes that need to be identified and quantitated. The IS should be within the same concentration range as for the calibration standards and at a fixed concentration. The following is a calibration curve for the IS mode:

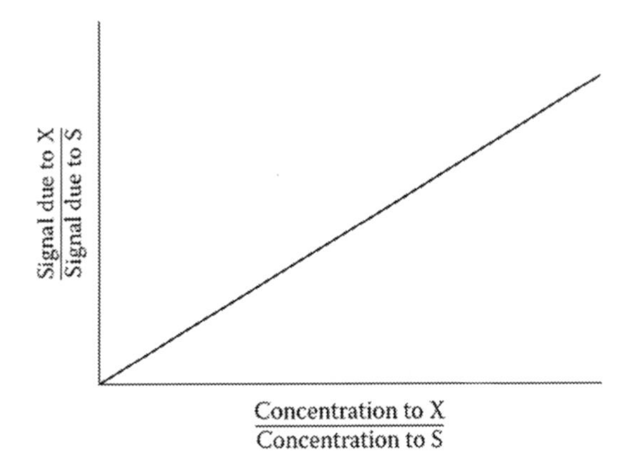

This exercise introduces the student to *reversed-phase solid-phase extraction* (RP-SPE) techniques. The *mini-LLE* method will also be implemented. The two methods will be compared. SPE in the reversed-phase (RP) mode of operation involves passing an aqueous sample over a previously conditioned sorbent that contains a chemically bonded silica gel held in place with polyethylene frits within a column configuration. A typical RP-SPE sequence follows:

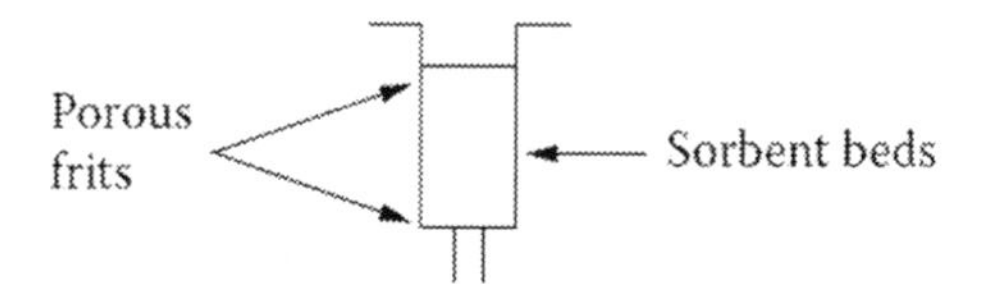

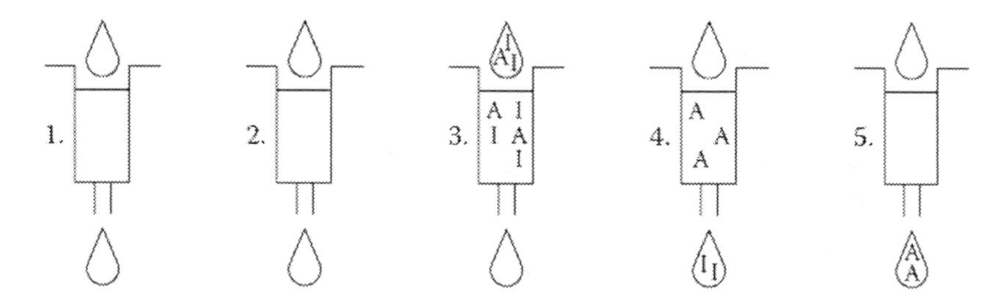

For RP-SPE, methanol is used to condition or wet the sorbent surface, thereby activating the octadecyl moiety and hence forcing it to be receptive to a van der Waals type of intermolecular interaction between the analyte and the C_{18} moiety. This phenomenon is shown below for the isolation of *n*-butyl phthalate on a C_{18} chemically bonded sorbent.

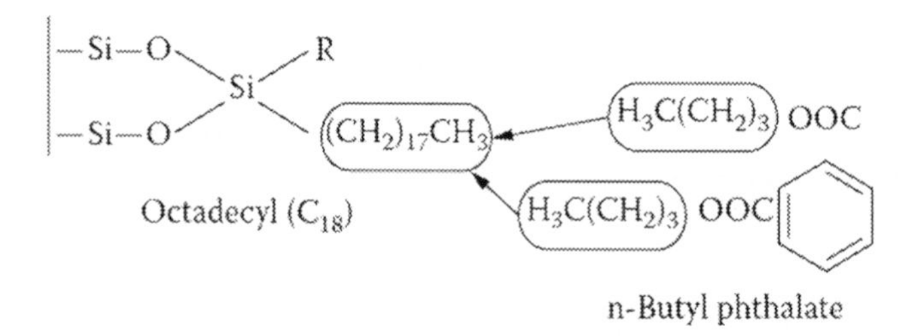

The OCs studied in this exercise are lindane, endrin, and methoxychlor. Lindane (γ-BHC) is synthesized via chlorination of benzene in the presence of ultraviolet light. This forms a mixture of BHC isomers that are identified as α, β, γ, δ, and ε. Selective crystallization isolates the γ isomer, whose aqueous solubility is 7.3 to 10.0 ppm and is the most soluble of the BHC isomers. Endrin is a member of the cyclodiene insecticides and is synthesized using Diels–Alder chemistry. Methoxychlor belongs to the *p, p'*-DDT category and structurally differs from DDT in substitution of a methoxy group in place of a chloro group *para* to the central carbon. Methoxychlor's aqueous solubility is 0.1 to 0.25 ppm and exceeds that of *p, p'*-DDT by a factor of 100. Molecular structures and correct organic nomenclature of these three representative OCs are shown below:

Gamma-BHC

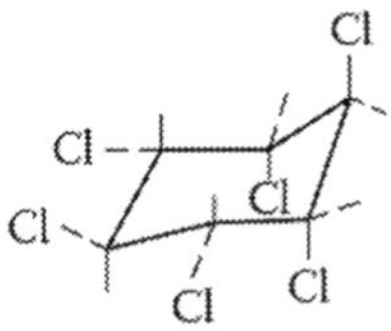

γ-1,2,3,4,5,6-Hexachlorocyclohexane

Endrin

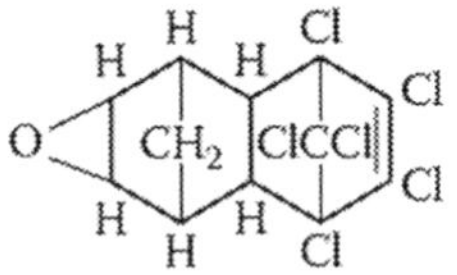

1,2,3,4,10,10-Hexachloro-exo-6,7-epoxy-1,4,4a,5,-6,7,8,8a-octahydro-1,4-endo, endo-5,8-dimethanonaphthalene

P,P'-Methoxychlor

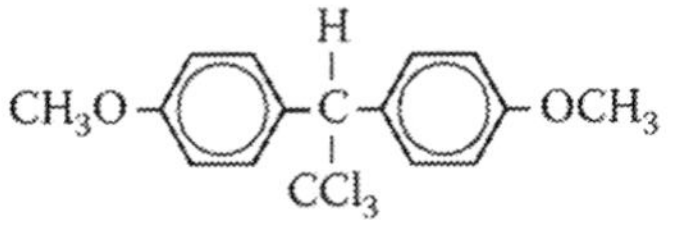

1,1,1-Trichloro-2, bis (p-methoxyphenyl) ethane

18.2 EXPERIMENTAL

Gas chromatograph that incorporates an electron-capture detector.

18.2.1 Preparation of Chemical Reagents

Note: all reagents used in this analytical method contain hazardous chemicals. Wear appropriate eye protection, gloves, and protective attire. The use of concentrated acids and bases should be done in the fume hood.

18.2.2 Chemicals/Reagents Needed per Group

One 1,000 ppm each of lindane, endrin, and methoxychlor stock standard solution dissolved in high-purity iso-octane. This is a solvent available in ultrahigh purity, which is an important requirement in trace environmental quantitative analysis. Look for the *pesticide-grade label* on the bottle of iso-octane to verify its purity.

One 20 ppm of an internal standard. Available candidates include 4-hydroxy-2,4,6-trichlorobiphenyl, 3,4,3',4'-tetrachlorobiphenyl, 1,2-dibromo-3-chloropropane, and β-BHC.

One vial containing approximately 30 mL of methanol for conditioning the RP-SPE sorbent.

One vial containing approximately 30 mL of high-purity iso-octane.

Molecular structures for the IS candidates are shown below:

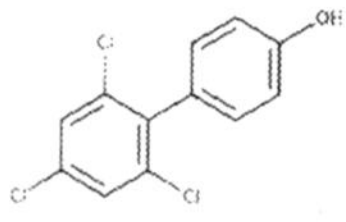

4-Hydroxy-2,4,6-trichlorobiphenyl 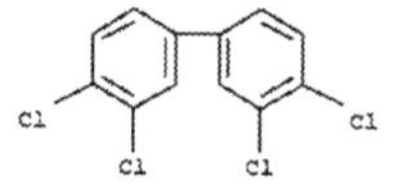 3,4,3',4'-tetachlorobiphenyl 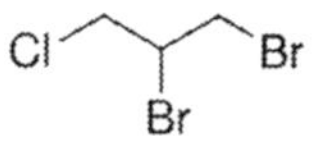 1,2-dibromo-3-chloropropane 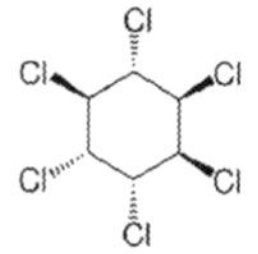 β-BHC

18.2.3 Preliminary Planning

Because there are two sample preparation methods to be implemented, assemble as a group at the beginning of the laboratory session and decide who does what. Once all results are obtained, the group should reassemble and share all analytical data.

18.2.4 Selection of a Suitable Internal Standard

The most appropriate IS needs to be selected from the above list of candidates. Consult with your laboratory instructor and proceed to inject one or more ISs and base your decision on an interpretation of the chromatogram.

18.2.5 Procedure for Calibration and Quantitation of the GC-ECD

1. Prepare the necessary primary and secondary dilution standards. The range of concentration levels for the unknowns is between 100 and 1,000 pg/μL (ppb). For example, take a 100 μL aliquot of the 1,000 ppm stock and add it to a 10 mL volumetric flask previously half filled with iso-octane. Adjust to the calibration mark and label "10 ppm L, E, M (iso-octane), primary dilution." A 1:10 dilution of this primary dilution standard gives a *secondary dilution*, which should be labeled "1 ppm L, E, M (iso-octane), secondary dilution." Use the secondary dilution to prepare a *series of working calibrations standards* that cover the range of concentrations in the ppb domain, as discussed above.
2. Prepare a set of working calibration standards to include an ICV that brackets the anticipated range for the unknowns. To each calibration standard, add 50 μL of 20 ppm IS so that the *concentration of IS in each calibration standard is identical at 1.0 ppm.*
3. Retrieve the method from Turbochrom or other equivalent software titled "LEMIS," which stands for *lindane, endrin, methoxychlor, internal standard* mode of calibration; allow sufficient instrument equilibration time. Write a sequence encompassing the calibration standards, ICV, and unknowns. Save the sequence as a file with e.g. the name "G#0317" (Group #, March 17). Begin to inject a 1 μL aliquot of each working standard. Initially inject iso-octane, then inject in the order of lowest to highest concentration level. This order is important because it prevents carryover from one standard to the next.

4. Update the calibration for the LEMIS method and check with your instructor as to the acceptability of the calibration. If found acceptable, proceed to the analysis once samples have been prepared using both extraction methods. *Be sure to add the same amount of IS to each unknown extract, as was done for the calibration standards.* Because the instrument has been calibrated and updated, the report will include an accurate readout of concentration in a tabular format.

18.2.6 Procedure for Performing Mini-LLE and RP-SPE

1. Place exactly 40 mL of unknown sample into a 42 mL vial and extract using 2 mL of iso-octane in a manner similar to that for the BTEX/THMs exercise. This time, however, add twice the amount of IS that you added for the preparation of the calibration standards so that the concentration of IS remains identical to that for all other standards and samples.
2. Place exactly 40 mL of unknown sample into the 70 mL SPE reservoir, which sits atop a previously conditioned C_{18} sorbent, according to specific instructions given to you by your laboratory instructor. Add distilled deionized water (DDI) to the reservoir so as to fill to near capacity. Pass the aqueous sample through the cartridge, which contains approximately 200 mg of C_{18} chemically bonded silica gel. Use a wash bottle that contains DDI to rinse the residual sample from both the reservoir and the cartridge. *Place a second SPE cartridge that is filled with anhydrous sodium sulfate beneath the sorbent cartridge.* The second SPE cartridge containing anhydrous Na_2SO_4 is used to remove residual water from the eluent. Into the manifold place a 1.0 mL volumetric flask as an eluent receiver and elute with two successive 500 μL aliquots of iso-octane. Add the same amount of IS as used for the calibration standards, then adjust to a final volume of 1.0 mL. Transfer to a separate container if necessary.
3. Inject a 1 μL aliquot of the sample extract that also contains the IS into the GC-ECD. At this point, the LEMIS method should have had its calibration updated.
4. Continue to make injections into the calibrated GC-ECD until all samples have been completed. You may want to make replicate injections of a given sample extract.

18.3 FOR THE REPORT

Include all calibration plots and calculate the correlation coefficient for the calibration plot. Calculate the precision and accuracy for the ICV, which should have been injected in triplicate. Report on the concentration of each unknown sample. Construct a table that shows the respective concentrations for the unknowns for each of the two methods. Recall that the final extract volume from mini-LLE was 2 mL, and that from RP-SPE was 1 mL. Take this into account when comparing the two methods. Which sample preparation method is preferable? Give reasons for your preference and support this with analytical data.

18.4 SUGGESTED READING

To develop this experiment, the author consulted the following resources:

Lee H., A. Chau, F. Kawahara. Organochlorine pesticides. In Chau A., B. Afghan Eds. *Analysis of Pesticides in Water.* Vol II. Boca Raton, FL: CRC Press, 1982. Chapter 1. This work is part of a three-volume series and is a good introduction to pesticide residue analysis even though it is outdated in the use of packed, instead of capillary GC columns.

Loconto P.R. Quantitating Toxaphene Parlar congeners in fish using large volume injection isotope dilution GC with electron-capture negative ion MS. *LC-GC North America*, 36(5): 320–328, 2018. The author's most recent contribution discusses how to develop and validate a new analytical method to quantitate this ubiquitous organochlorine legacy pesticide down to low ppt concentration levels. Archives of back issues LC-GC are available online.

Methods for Organic Chemical Analysis of Municipal and Industrial Wastewaters. EPA 600/4-82-057. Cincinnati, OH: Environmental Monitoring and Support Laboratory, July 1982.

Perry J. *Introduction to Analytical Gas Chromatography.* New York: Marcel Dekker, 1981, pp. 164–175. One of the better discussions of Lovelock's development of the electron-capture detector.

Sawyer D., W. Heineman, J. Beebe. *Chemistry Experiments for Instrumental Methods.* New York: Wiley, 1984, pp. 321–343.

SPE Sample Preparation. Phillipsburg, NJ: J.T. Baker Chemical Co., 1984. p. 8.

Thurman E., M. Mills. *Solid-Phase Extraction: Principles and Practice.* New York: Wiley-Interscience, 1998. Following an overview and theoretical background for SPE sorption and isolation, the authors discuss method development practices using SPE. This topic is followed by introducing the principles of reversed-phase, normal-phase, and ion-exchange SPE. These topics are followed by applications of SPE to environmental, pharmaceutical, food, and natural product samples.

19 Determination of Priority Pollutant Polycyclic Aromatic Hydrocarbons (PAHs) in Contaminated Soil Using RP-HPLC-PDA with Wavelength Programming

19.1 BACKGROUND AND SUMMARY OF METHOD

In 1979, the EPA proposed Method 610, which, if properly implemented, would determine the 16 priority pollutant PAHs in municipal and industrial discharges. The method was designed to be used to meet the monitoring requirements of the National Pollutant Discharge Elimination System (NPDES). The assumption used was that a high expectation of finding some, if not all, of the PAHs was likely. The method incorporated packed-column GC in addition to HPLC, and because of the inherent limitation of packed columns, they were unable to chromatographically resolve four pairs of compounds (e.g. anthracene from phenanthrene). Because RP-HPLC could separate all 16 PAHs, it become the method of choice. The method involved extracting a 1 L sample of wastewater using methylene chloride, using *Kuderna–Danish evaporative concentrators* to reduce the volume of solvent, cleaning up the extract using a silica gel micro-column, and implementing a solvent exchange to acetonitrile prior to making an injection into an HPLC system. The method requires that a UV absorption detector and a fluorescence emission detector be connected in series to the column outlet. This affords maximum detection sensitivity because some PAHs (e.g. naphthalene, phenanthrene, fluoranthene, among others) are much more sensitive when detected by fluorescence emission when compared to detection by UV absorption.

In most laboratories today, PAHs are routinely monitored under EPA Method 8270 which incorporates a WCOT GC column and includes the majority of neutrals under the base, neutral, acid (BNAs) designation of the method. This is a liquid–liquid extraction method with a quantitative determination utilizing gas chromatography-mass spectrometry (GC-MS). Careful changes in pH of the aqueous phase enable a selective extraction of bases and neutrals from acidic compounds. Examples of

DOI: 10.1201/9781003260707-19

priority pollutant organic bases include *aniline and substituted anilines*. Examples of priority pollutant organic acids include *phenol and substituted phenols*. The most popular method of recent years has been EPA Method 525, which incorporates RP-SPE techniques and is applicable to PAHs in drinking water.

The most common UV wavelength, λ, for use with aromatic organic compounds is generally 254 nm because almost all molecules that incorporate the benzene ring in their structure absorb at this wavelength. This wavelength may or may not be the most sensitive wavelength for most PAHs.

RP-HPLC chromatograms for the 16 priority pollutant PAHs in a reference standard mixture (top) and from a soil extract (bottom) are shown below. In the lower chromatogram of each figure, λ was held fixed at 255 nm, whereas for the upper chromatogram of each figure, λ was changed during the run so as to demonstrate how the wavelength influences peak height. The wavelength-programmed HPLC chromatogram shows much less background absorbance and hence increased sensitivity. This information should be used in developing the wavelength-programmed HPLC method.

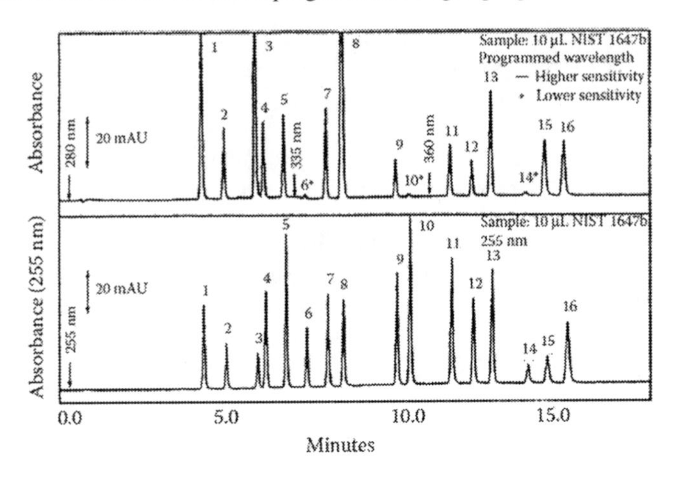

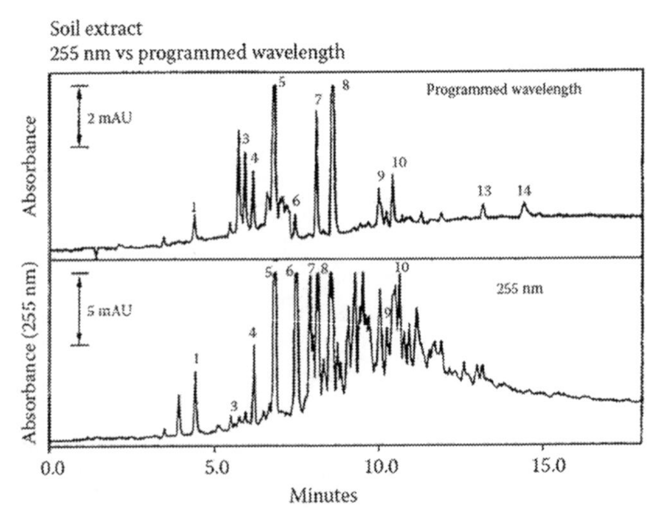

19.2 OF WHAT VALUE IS THIS EXPERIMENT?

The exercise affords the student an opportunity to build a new HPLC method using the chromatography data-handling software. The method will also incorporate the concept of wavelength programming, whose objective is to maximize detector sensitivity for a given analyte and which can only be performed using a photodiode array (PDA) detector and accompanying digital electronics. The following table summarizes the detection limits for each PAH in terms of nanograms (ng) injected for $\lambda = 255$ nm and for $\lambda = 280$ nm and for UV programming during the chromatographic run:

Sensitivity and Linearity Data for UV Absorption Detection

No.	PAH	$\lambda = 255$ nm (ng)	$\lambda = 280$ nm (ng)	$\lambda =$ Programmed (ng)
1	Naphthalene	0.6	0.7	0.7 (280 nm)
2	Acenaphthylene	0.9	1.9	1.9
3	Acenaphthene	1.42	0.60	0.60
4	Fluorene	0.13	0.53	0.53
5	Phenanthrene	0.06	0.36	0.36
6	Anthracene	0.03	2	1.2 (335 nm)
7	Fluoranthene	0.22	0.24	0.45
8	Pyrene	0.25	1.1	0.07
9	Benz(a)anthracene	0.09	0.08	0.5
10	Chrysene	0.06	0.41	6.0
11	Benzo(f)fluoranthene	0.09	0.23	0.4 (360 nm)
12	Benzo(k)fluoranthene	0.14	0.31	0.6
13	Benzo(a)pyrene	0.11	0.2	0.2
14	Dibenz(a, h)anthracene	0.45	0.14	4
15	Benzo(g, h, i)perylene	0.32	0.32	0.3
16	Indeno(1, 2, 3-c, d)pyrene	0.16	0.38	0.35

19.3 EXPERIMENT

A high-performance liquid chromatograph (HPLC) that incorporates a UV absorption detector such as a photodiode array under reversed-phase conditions is necessary to perform this experiment.

19.3.1 Preparation of Chemical Reagents

Note: all reagents used in this analytical method contain hazardous chemicals. Wear appropriate eye protection, gloves, and protective attire. The use of concentrated acids and bases should be done in the fume hood.

19.3.2 Accessories to Be Used with the HPLC per Group

One HPLC syringe. This syringe incorporates a blunt end; the use of a beveled-end GC syringe would damage inner seals to the Rheodyne injector.
One 16-component PAH standard. Check the label for concentration values.

19.3.3 Procedure

Be sure to record your observations in your laboratory notebook.

19.3.3.1 Creating the Wavelength Program Method

Again, you will first find the HPLC instrument in the off position; use "hands on" to activate the instrument and allow at least 15 min for the detector to warm up and stabilize. Ask your laboratory instructor for assistance if necessary. Observe the variability in baseline absorbance. Absorbance should not vary much above a ΔA = 0.0100. Significant variability is most often due to trapped air bubbles because of insufficient degassing of the mobile phase. Inform your instructor if this baseline absorbance variation is significant.

Once the baseline is stable, retrieve the method titled "PAH255" and download it. This method is one previously created by the instructional staff and is a fixed wavelength (λ at 255 nm). Fill the 5 μL injection loop with the PAH standard and observe the chromatogram. The method separates the PAHs based on gradient elution. The method incorporates a one-point calibration.

Use the above tabular information and modify this method to incorporate wavelength programming as discussed earlier. Save the modified method as "PAHWP," where WP stands for "wavelength programmed." Ask your laboratory instructor for assistance in developing this software capability. Fill the 5 μL injection loop with the PAH standard. Using the "chromatograms" section in the main menu, proceed to retrieve both HPLC chromatograms that you just generated. Use the overlay capability to compare both chromatograms and print the overlay. Update the one-point calibration with this standard. You should now have a new method with an updated calibration prior to injecting the extract from the soil discussed below.

19.3.3.2 Extraction Procedure for Soil

Weigh approximately 2.0 g of contaminated soil into a 50 or 125 mL glass beaker. Add 20 mL of methylene chloride and use a glass stirring rod to facilitate mixing. Let the contents of the mixture stand for at least 10 min. Decant the extract into a second beaker. It may be necessary to filter this extract if particulates become a problem. This will depend on the type of sample. Pipette 1.0 mL of the methylene chloride extract into a clean, dry 10 mL volumetric flask. Adjust to the calibration mark with *acetonitrile*. Fill the injection loop with this diluted extract. It may be necessary to use a 0.45 μm syringe filter to remove particulates from the diluted extract. Fill the HPLC syringe with about five times the loop volume to ensure a reproducible injection volume. The peak area that is found refers to the concentration of a given PAH in the diluted extract. You will be given assistance on how to allow Turbochrom® (PerkinElmer) to calculate the concentration of each PAH in the original contaminated soil. If time permits, make a second injection of the diluted extract. *Discard the excess methylene chloride extract and CH_2Cl_2/ACN diluted extract into a waste receptacle when finished.*

19.3.3.3 Calculation of the # ppm of Each PAH in Contaminated Soil

Let us assume that upon injection of the diluted soil extract, a concentration of 225 ppm dibenzo(*a*, *h*)anthracene in the diluted soil extract was obtained based on a correctly calibrated instrument.

What would the original concentration of dibenzo (*a*, *h*) anthracene be in the contaminated soil?

Please study the series of calculations shown below:

- 225 ppm means 225 μg/mL of diluted soil extract.
- Thus, 225 × 10 = 2,250 μg/mL in the original 20 mL of extract before dilution.
- One says that the dilution factor DF is 10.
- (20 mL extract) (2,250 μg/mL dibenzo[*a*, *h*]anthracene) = 45,000 μg total from 2 g of soil.
- 45,000 μg total/2.0 g soil = 22,500 μg/g or ppm dibenzo(*a*, *h*)anthracene in contaminated soil.

Upon properly completing the sequence file within the Turbochrom® software, the final result, 22,500 ppm, will be directly obtained in the "peak report" for that sample.

19.4 FOR THE REPORT

Include the overlay comparison and calibration results and list the concentration of each PAH in the contaminated soil sample. If a second sample result is available, estimate the precision of the method. One usually needs at least three replicate results to begin to use statistics to calculate an acceptable standard deviation. Comment on the advantage of using a PDA to increase sensitivity.

Address the following (refer to the table below on PAHs to assist in your answers):

1. Explain the elution order for the 16 PAHs using chemical principles.
2. The method detection limit using a UV absorption detector for some of the 16 priority pollutant PAHs could be improved if a different detector could be used. Explain.
3. Explain why this method is considered "quick." Are there limitations to the use of quick methods, and if so, what are some of these?

Some representative PAHs with corresponding physicochemical properties are shown below and include: molecular weight (MW), molecular formula, molecular structure, aqueous solubility (#mg/L), and the logarithm of the octanol-water partition coefficient:

Compound	Abbreviation	MW	Molecular Formula	Molecular Structure	Aqueous Solubility (#mg/L)	Log (Kow)
Naphthalene	NA	128	$C^{10}H^{8}$		31.7	3.36
Acenaphthylene	ACY	152	C12H8		16.1	3.94
Acenaphthene	ACE	154	$C_{12}H_{10}$		3.93	4.03
Fluorene	FLE	166	$C_{13}H_{10}$		1.98	4.47
Phenanthrene	PH	178	$C_{14}C_{10}$		1.29	4.57
Anthracene	AN	178	$C_{14}H_{10}$		0.073	4.54
Fluoranthene	FLA	202	$C_{16}H_{10}$		0.260	5.22

(Continued)

Compound	Abbreviation	MW	Molecular Formula	Molecular Structure	Aqueous Solubility (#mg/L)	Log (Kow)
Pyrene	PY	202	$C_{16}H_{10}$	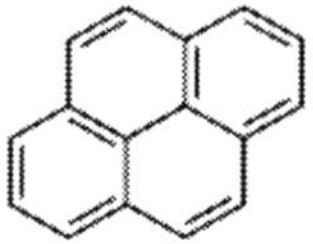	0.135	5.18
Triphenylene	TRP	228	$C_{18}H_{12}$	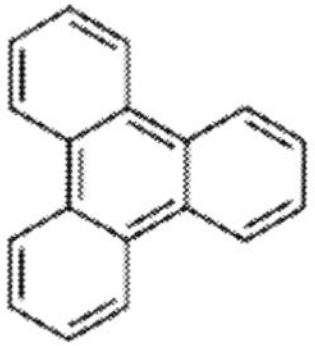	0.043	5.45
Benz(a)anthracene	BaA	228	$C_{18}H_{12}$	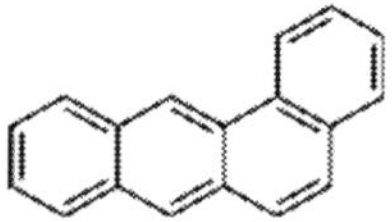	0.014	5.91
Chrysene	CHR	228	$C_{18}H_{12}$	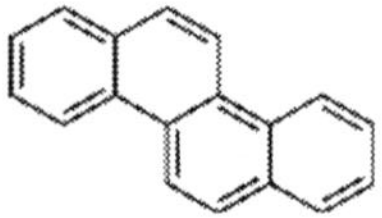	0.002	5.91

19.5 SUGGESTED READING

To develop this experiment, the author consulted the following resources:

Determination of organic compounds in drinking water by liquid–solid extraction and capillary gas chromatography/mass spectrometry. In *Methods for the Determination of Organic Compounds in Drinking Water*, Method 525, EPA/600/4–88/039. Cincinnati: EMSL, 1988.

Dong M. et al. A quick-turnaround HPLC method for PAHs in soil, water and wastewater oil. *LC-GC* 11: 802–810, 1993. Archives of back issues of LC-GC are available online.

Method 8270E. Gas chromatography–mass spectrometry for semi-volatile organics: Capillary column technique. In *Test Methods for Evaluating Solid Wastes*, Washington, DC: Office of Solid Waste, EPA, SW 846 Update VI, Revision 8, June 2018.

Polynuclear aromatic hydrocarbons. *Method 610 Federal Register* 233: 69514–69517, 1979.

Sawyer D., W. Heineman, J. Beebe. *Chemistry Experiments for Instrumental Methods.* New York: Wiley, 1984, pp. 344–360.

Snyder L., J. Kirkland. *Introduction to Modern Liquid Chromatography.* 2nd edn. New York: Wiley, 1979.

Snyder L., J. Glajch, J. Kirkland. *Practical HPLC Method Development.* New York: Wiley, 1988.

Snyder L., J. Kirkland. J. Glajch. *Practical HPLC Method Development.* 2nd edn. New York: Wiley, 1997.

Stoll D., T. Lauer. Effects of buffer capacity in reversed-phase liquid chromatography, Part 1; Relationship between the sample- and mobile phase- buffers. *LC-GC North America* 38: 10–15, 2020. Exemplifies current thinking in RP-HPLC research and development. Archives of back issues of LC-GC are available online.

20 How to Set up and Operate an Ion Chromatograph

Students will notice that on the syllabus for the laboratory course for the first experiment titled: "Introduction to pH Measurement: Estimating the Degree of Purity of Snow," it was stated that an ion chromatograph was available, to be operated by the staff, for students to analyze their snow samples in an attempt to measure impurities. Below is an additional experiment/exercise that provides an opportunity for students (perhaps for extra credit) to learn to operate an additional analytical instrument known as an ion chromatograph.

20.1 DETERMINATION OF INORGANIC ANIONS USING ION CHROMATOGRAPHY (IC): ANION EXCHANGE IC WITH SUPPRESSED CONDUCTIVITY DETECTION

20.1.1 BACKGROUND

An aqueous sample obtained from the environment may be expected to contain dissolved inorganic salts. The concentration of these salts can be found by preparing and injecting aqueous samples into a properly installed and optimized ion chromatograph. The concentration of the most common inorganic anions and corresponding cations can be quantitatively determined. The instrumentation available in our laboratory is that originally manufactured by the Dionex Corporation. Dionex is today part of Thermo Fisher Scientific Corporation.

Ion chromatography (IC) is a low- to moderate-pressure liquid chromatographic (LC) technique and should be clearly distinguished from that of high-pressure LC (HPLC). IC as a determinative technique has been developed to *separate and detect both cations and anions*. The instrument available to the student utilizes anion exchange IC with suppressed conductivity detection. This technique can separate the *common inorganic anions* found in aqueous environmental samples—fluoride (F^-), chloride (Cl^-), bromide (Br^-), nitrite (NO_2^-), nitrate (NO_3^-), phosphate (PO_4^{3-}), and sulfate (SO_4^{2-})—under the IC conditions used here. A second set of related anions of environmental interest has emerged in recent years, and these are collectively called *inorganic disinfection by-products* and include bromate, bromide, chlorite, and chlorate. EPA Method 300.1 has been recently revised, and this method addresses both sets of analytes. Prior to the development of the micromembrane

DOI: 10.1201/9781003260707-20

suppressor, a suppressor column was used to reduce the background conductivity of the mobile-phase eluent. The following is a schematic drawing of the classical ion chromatograph that includes the original suppressor column:

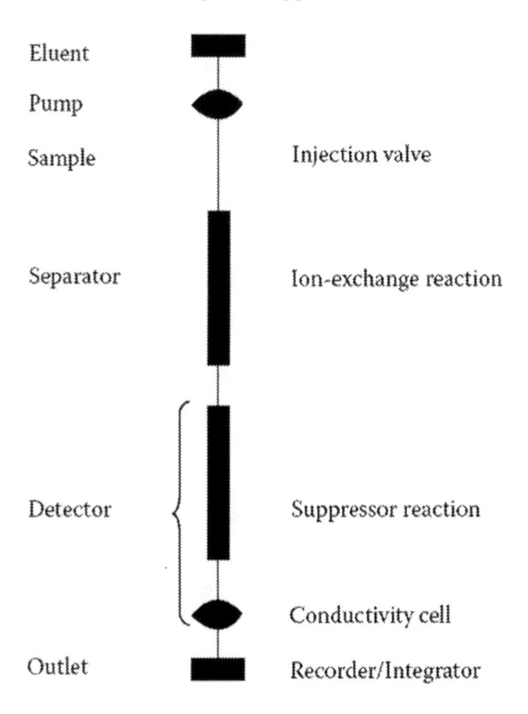

For anion exchange IC, the eluent must be a moderate to strong base (e.g. carbonate or hydroxide). Carbonate (CO_3^{2-}) or hydroxide (OH^-) ions *displace the analyte anion* through the anion exchange resin in the separator column. The separator must contain a low-capacity anion exchange resin so that the analyte ions can make it through the column in a reasonable length of time after injection. Because the conductivity detector responds to all ions, a strong signal due to the eluent would be observed, thus "swamping out" the contribution due to the analyte anions. These eluent ions can be chemically removed while the analyte ions elute from the suppressor in a low-conductivity background. This is done in the suppressor, and the conductivity of the eluent is said to be chemically suppressed. The suppressor and conductivity cell comprise the IC detector, as shown above. The suppressor must contain a cation exchange substrate whereby H^+ from the regenerant migrates across the membrane (which is itself a cation exchanger) and neutralizes the carbonate or hydroxide to form neutral carbonic acid or water. The micromembrane suppressor requires a continuous supply of regenerate solution that consists of 0.025 *N* sulfuric acid. The regenerant used in our laboratory utilizes mid-1980s technology and consists of a reservoir that contains the dilute H_2SO_4. This solution is made to flow into and out of the micromembrane suppressor by means of positive air pressure. More contemporary designs self-generate the H_2SO_4 electrochemically.

A typical ion chromatogram for separation and detection of a reference standard that contains all seven *common* anions follows:

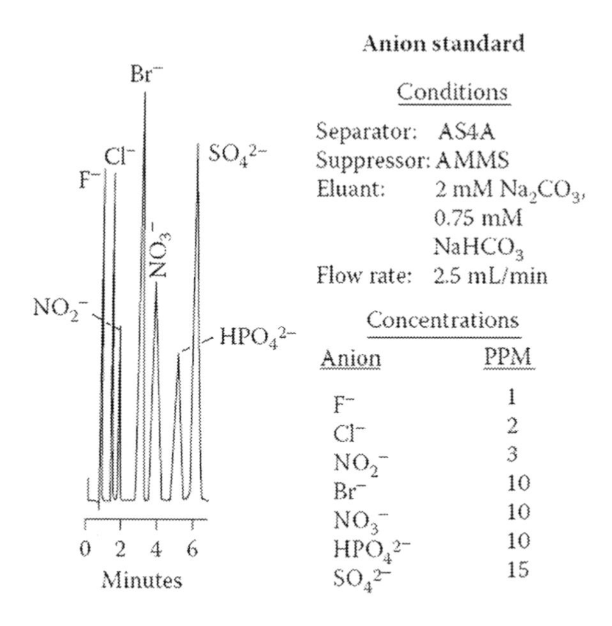

20.2 HOW DO I OPERATE THE ION CHROMATOGRAPH?

To operate the ion chromatograph, review the operator's manual. Alternatively, your instructor may have a written procedure. For purposes of illustration, we list below a procedure used in our laboratory. A Model 2000® (Dionex) instrument interfaced to a PC via a 900® (PE Nelson) interface is available:

1. Turn the compressed air valve on and adjust the main pressure gauge to 80 psi. This provides a head pressure to the eluent reservoir.
2. Adjust the small pressure gauge located adjacent to the ion chromatograph to approximately 10 psi. This provides a head pressure to the regenerant reservoir.
3. Turn on the power to the chromatography module/SP via a switch located in the rear of the module. This will start the single-piston reciprocating pump on the Model 2000® ion chromatograph.
4. Turn the conductivity detector cell display to "on" and monitor the output. This is located on the detector module. A reading between 10 and 20 μS (micro-siemens) with a tolerance of < 1 μS represents a stable baseline. This implies that the ion chromatograph is sufficiently equilibrated across the micro-membrane suppressor. At times, the regenerate flow rate may need to be increased or decreased by adjusting the head pressure. If regenerate is not flowing through the micro-membrane suppressor, the conductivity value will "skyrocket."

5. At the PC workstation, retrieve the appropriate method from the Turbochrom® (PE Nelson) software. The document "Anion.mth" is available if a more specific method has not been created.
6. Either create a sequence or use the method under "setup" and proceed to download the method. This enables the instrument and PC to communicate.
7. Once the workstation PC reaches a ready status, press inject to remove the lit LED on the Dionex Automation Interface. Fill the 50 μL injection loop, which is located at the sample port on the chromatography/SP, with a filtered, aqueous sample. Press the "inject" and immediately proceed to the 900® (PE Nelson) interface box and press "start."
8. After the chromatographic run is complete, repeat step 7 by first depressing the "inject" to remove the LED.

20.3 IS THERE A NEED FOR SAMPLE PREP?

Yes and no. Aqueous samples that are free of suspended matter and contain dissolved inorganic salts are the only type of sample matrix that is suitable for direct injection into the IC. Wastewater samples that contain suspended solids must be filtered prior to injection into the IC. *Wastewater samples* that have a dissolved organic content (i.e. an appreciable total organic carbon [TOC]) level should be passed through a previously conditioned reversed-phase solid-phase extraction (RP-SPE) cartridge to attempt to remove the dissolved organic matter prior to injection into the IC. Keep in mind that these RP-SPE cartridges have a finite capacity. If this capacity is exceeded, contaminants will merely pass through. If *samples come from a bioreactor*, proteins and other high-molecular-weight solutes must also be removed prior to injection into the IC.

After implementing the operations procedure, the eluent should be pumping through the separator column and the micro-membrane suppressor, while the regenerant should be flowing in the opposite direction to the eluent flow, through the suppressor under building-supplied compressed air.

Also included in this experiment are specific procedures to prepare the bicarbonate/carbonate eluent, the preparation of the mixed anion reference standards, and weights of various salts to be used to prepare stock reference standard solution for all analytes of interest.

20.4 HOW DO I PREPARE A REFERENCE STOCK STANDARD FOR EACH ANION?

Note: all reagents used in this analytical method contain hazardous chemicals. Wear appropriate eye protection, gloves, and protective attire. The use of concentrated acids and bases should be done in the fume hood.

The weight of each salt that can be used to prepare a *1,000 ppm reference stock standard as the anion portion of the salt* without any reference to a cation is listed in the following table:

Salt	Grams to Dissolve to Prepare 1 L of a 1,000 ppm as X
NaF	2.210
NaCl	1.646
KBr	1.488
$NaNO_2$	1.500
$NaNO_3$	1.371
$K2SO_4$	1.814
$KBrO_3$	1.315
$KClO_3$	1.467
$NaClO_2$	1.341

Note: where X = F, Cl, NO_2, etc.

20.5 HOW DO I PREPARE THE BICARBONATE/CARBONATE ELUENT FROM SCRATCH?

Dissolve 0.571 g of $NaHCO_3$ and 0.763 g of Na_2CO_3 in approximately 250 mL of distilled deionized water (DDI). Transfer this solution to a 200 mL graduated cylinder. Adjust to the mark with DDI and transfer this solution to the IC reservoir. Add 2,000 mL more of DDI to the IC reservoir for a total of 4 L. Label the IC reservoir as "1.7 m*M* HCO_3/1.8 m*M* CO_3" eluent. This eluent is used with either an AS4A® (Dionex) or equivalent anion-exchange IC separator column.

20.6 HOW DO I PREPARE A MIXED ANION STOCK STANDARD FOR IC?

Certified reference stock standard solutions for each of the anions at *1,000 ppm each anion* must be available. The following table outlines one approach to prepare a *mixed-stock reference standard*:

1,000 ppm Stock (#mL)	Anion	Concentration (ppm)
3.0	Chloride	3.0
1.0	Nitrite	1.0
10.0	Bromide	10
2.0	Nitrate	2.0
20.0	Phosphate	20.0
10.0	Sulfate	10.0

A *clean, dry* 1,000 mL volumetric flask must be used. Pipette the indicated aliquot into a flask that is approximately half-filled with DDI.

20.7 HOW DO I PREPARE A FOUR-LEVEL SET OF CALIBRATION STANDARDS FOR IC?

Using the *mixed reference stock solution* (presented earlier), proceed using the following table as a guide to prepare a set of working calibration standards:

Mixed-Stock Reference (mL)	V (mL)	#ppmC$^-$	#ppmNO_2^-	#ppmBr$^-$	#ppmNO_3^-	#ppmHPO_4^-	#ppmSO_4^{2-}
5.0 (Cal Std #1)	25	0.6	0.2	2.0	0.4	4.0	2.0
10.0 (Cal Std #2)	25	1.2	0.4	4.0	0.8	8.0	4.0
20.0 (Cal Std #3)	25	2.4	0.8	8.0	1.6	16.0	8.0
Neat (Cal Std #4)	—	3.0	1.0	10.0	2.0	20.0	10.0

To prepare a set of four calibration standards, let's consider how to prepare, for example, *Calibration Standard* (Cal Std #1): pipette 5.0 mL of the mixed *anion stock reference* solution and transfer this aliquot to a clean, dry 25 mL glass volumetric flask, then dilute to the calibration mark on the volumetric flask. This yields Cal Std #1 whose concentration of chloride ion Cl^- – *is 0.6 ppm*, whose concentration of *$NO2^-$ is 0.2 ppm*, whose concentration of *Br^- is 2.0 ppm*, whose concentration of NO_3^- is 0.4 ppm, whose concentration of $HPO4^{2-}$ is 4.0 ppm and whose concentration of *$SO4^{2-}$– is 2.0 ppm*. Repeat for Cal Std #2 and for Cal Std #3. For Cal Std #4 inject the *mixed anion stock reference* solution directly into the ion chromatograph.

20.8 WHAT DOES THE DATA LOOK LIKE?

Two ion chromatograms that were run on the Model 2000® Dionex instrument are included in this experiment, shown in the following figure. A separation of six of the seven common anions (PO_4^{3-} not included) only 5 min after the instrument was turned on is shown in the first *ion chromatogram* below. Note the gradual rise in the baseline during the development of the chromatogram. This same reference standard of six anions was injected long after the baseline had stabilized. The baseline stability is shown in the second ion chromatogram. A stable baseline is essential for reproducible peak areas, and hence leads to good precision and accuracy for the trace quantitative determination of the common inorganic anions.

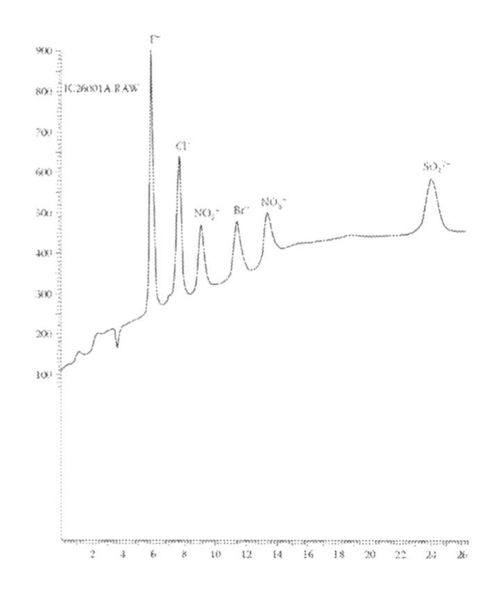

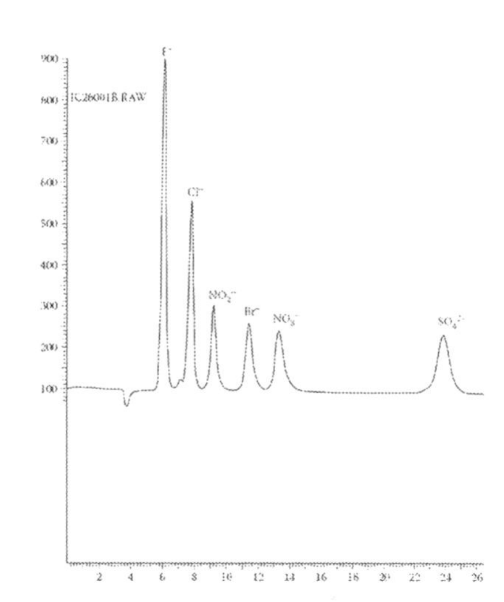

20.9 SUGGESTED READING

To develop this experiment, the author consulted the following resources:

Loconto P.R., N. Hussain Automated coupled ion exclusion: Ion chromatography for the determination of trace anions in fermentation broth. *Journal of Chromatographic Science* 33: 75–81, 1995. This paper demonstrates how IC and ion exclusion chromatography can be coupled together whereby inorganic anions such as Cl–, NO3–, PO43–, and SO42– can be detected and quantitated directly in a complex sample matrix such as fermentation broth.

Shpigun O., Y. Zolotov. *Ion Chromatography in Water Analysis.* Chichester, West Sussex, UK: Ellis Horwood Limited, 1988. A well-presented treatment of theory and practice of IC. Includes an excellent discussion of membrane suppression techniques.

Small H., T. Stevens, W. Baumann. *Anal Chem* 47: 1801–1809, 1975. This paper describes the principles of suppressed IC developed at the Dow Chemical Co. These concepts laid the foundation for the Dionex Corporation. Today suppressed IC technology and its numerous advances are part of Thermo Fisher Scientific Corporation.

Weiss J. *Handbook of Ion Chromatography*, E.L. Johnson ed. Sunnyvale, CA: Dionex Corporation, 1986. There was no better treatment of the subject at that time.

Index

A

B

C

D

E

F

Printed in the United States
by Baker & Taylor Publisher Services